FESTOOL® ESSENTIALS

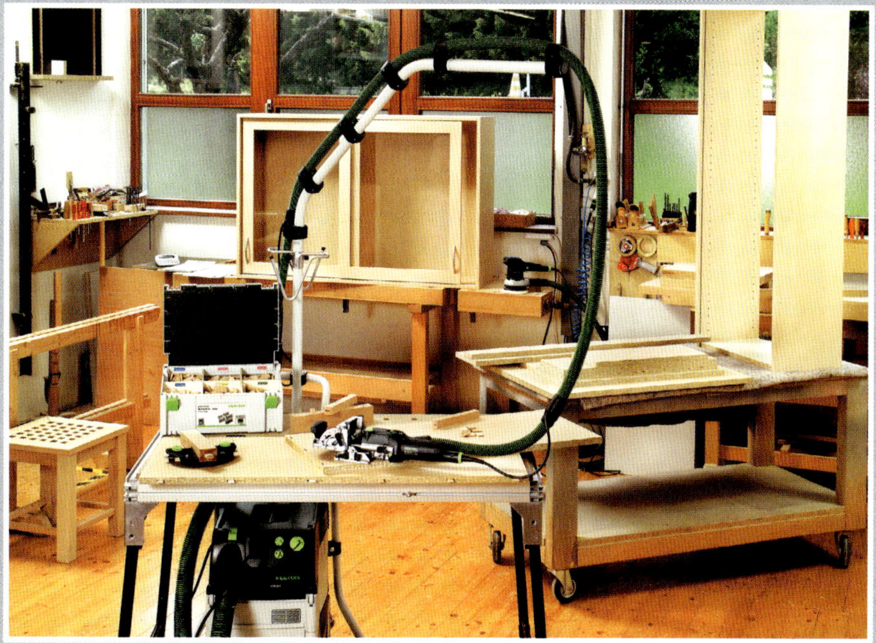

The DOMINO® DF 500Q Joiner

With CT Dust Extraction System

I0808823

Schiffer® Publishing Ltd

4880 Lower Valley Road Atglen, Pennsylvania 19310

Schiffer Books are available at special discounts for bulk purchases for sales promotions or premiums. Special editions, including personalized covers, corporate imprints, and excerpts can be created in large quantities for special needs. For more information contact the publisher:

Published by Schiffer Publishing Ltd.
4880 Lower Valley Road
Atglen, PA 19310
Phone: (610) 593-1777; Fax: (610) 593-2002
E-mail: Info@schifferbooks.com

For the largest selection of fine reference books on this and related subjects, please visit our web site at **www.schifferbooks.com**
We are always looking for people to write books on new and related subjects. If you have an idea for a book please contact us at the above address.

This book may be purchased from the publisher.
Include $5.00 for shipping.
Please try your bookstore first.
You may write for a free catalog.

In Europe, Schiffer books are distributed by
Bushwood Books
6 Marksbury Ave.
Kew Gardens
Surrey TW9 4JF England
Phone: 44 (0) 20 8392-8585; Fax: 44 (0) 20 8392-9876
E-mail: info@bushwoodbooks.co.uk
Website: www.bushwoodbooks.co.uk
Free postage in the U.K., Europe; air mail at cost.

Library of Congress Control Number: 2008928297

Covers and book designed by: Bruce Waters
Type set in Huminist 521

ISBN: 978-0-7643-3104-6
Printed in China

The Domino® DF 500 Q Joiner is the latest in long line of innovations by Festool®. Founded in 1925 by Albert Fezer and Gottlieb Stoll in southern Germany, the company has been known by many names, Fezer & Stoll, Festo, Festo Tooltechnic, and, beginning in 2000, Festool. Whatever the name, the mission has always been the same: to provide expert craftspeople with the finest tools available to make their work more efficient, easier, and better. From the beginning they have committed themselves to understanding the needs of the craftsperson in an everchanging workplace. They have addressed those needs with innovation, quality, and performance that is unsurpassed in the industry. The result is captured in their three word credo: "Faster, Easier, Smarter."

Today the name Festool stands for an integrated system of tools engineered for ease of use and reliability, to increase the accuracy, creativity, and productivity of the workplace. Each tool in designed to be used with the powerful dust extraction system, which is covered here in great detail.

Used with Festool plunge saws and routers, the Festool guide and work table systems allow for unsurpassed accuracy, convenience, and flexibility.

The Domino® DF 500 Q Joiner was introduced to the U.S. market in 2007. It provides a revolutionary twist on the old mortise and tenon method of joinery, maintaining its strength while making it easy and fast with the latest technological innovations. It accomplishes a degree of accuracy and versatility that other joining techniques, including biscuits and dowels, cannot approach. This tool, with its available accessories, allows the user to make strong, solid joints with a minimal amount of effort, even in complex situations.

Flexible enough to make great panel, frame, face, and even beveled joints, the unique cutting action and adjustable hole widths make loose tenon joinery faster, easier, and more accurate than ever before.

Aligning boards is as simple as making the first cut, and then using the optional cross-stop with aligning pins to index off of the previous cut for perfectly spaced mortises. Mortises can easily be made in thin or round stock using the optional trim stop or handrail fence.

What makes this happen is an innovative cutting method, a simple indexing system, and adjustable mortise sizing. The Domino® tenon is designed to fit precisely into the mortise created by the tool. It is 100% solid beech or sipo mahogany (for outdoor use). Engineered to be totally rotation-free, they distribute glue evenly through the joint using glue expansion slots designed into their surface. This ensures to ensure a good, strong joint. Far stronger than either biscuits or dowels, the Domino® system tenons will soon become your preferred method for joinery.

The intention of this new book is to be a complete guide to the Domino® system, including the dust extraction system with which it is designed to work. The features of these systems are explained in a clear, yet detailed way. With each step illustrated with a full color photograph, the basic joining techniques are explored, making this a handy reference to the remarkable Festool System.

Main Features of the Domino® System

• **Patented routing action.** The spiral cutter rotates and oscillates to create perfect mortises.

• **Hole width adjustment.** Adjusting the width of the mortise allows for easier alignment when joining panels.

• **Positioning.** With an adjustable fence angle, retractable stop pins, adjustable cutter height with step gauge, and precise mortise depth stop, the Domino® joiner enables precise mortise placement and spacing.

• **Pivoting fence.** Fence creates angled mortises from 0-90° with positive stops at 22.5°, 45°, 67.5°.

• **Depth control.** Five mortising depths allow for quick, precise adjustment to the size of the tenon being used.

• **Stepped height stop.** 7 accurate step positions make adjusting the fence height for different material thicknesses easy and fast.

• **Dust extraction.** When the Domino is used in combination with a dust extractor, the mortises are clean and ready for the next step. Good dust extraction greatly extends cutter life.

• **Carbide tip spiral bit.** Durable, long-lasting bit withstands 5,000-14,000 joints depending on material.

• **Rapid bit change.** Change a bit in under 30 seconds to match desired tenon size.

Specifications

Model Name	DOMINO DF 500 Q	DOMINO DF 500 Q Set
Item	574 258	574 283
Standard package includes		
DOMINO cutter D 5	•	•
Support bracket	•	•
Wrench	•	•
Trim stop		•
Cross stop		•
Plug-it power cord	•	•
Systainer	SYS 2	SYS 2
Specifications		
Power consumption	420 watts	420 watts
Cutter spindle speed	24,300 rpm	24,300 rpm
Routing depth stop	12/15/20/25/28 mm	12/15/20/25/28 mm
DOMINO slot cutter diameter	5/6/8/10 mm	5/6/8/10 mm
Routing height adjuster with scale	5-30 mm	5-30 mm
Movable stepped stop	16/20/22/25/28/36/40 mm	16/20/22/25/28/36/40 mm
Miter routing	0-90°; stop positions at 0°, 22.5°, 45°, 67.5°, 90°	0-90°; stop positions at 0°, 22.5°, 45°, 67.5°, 90°
Weight	7 lbs. (3.2 kg)	7 lbs. (3.2 kg)

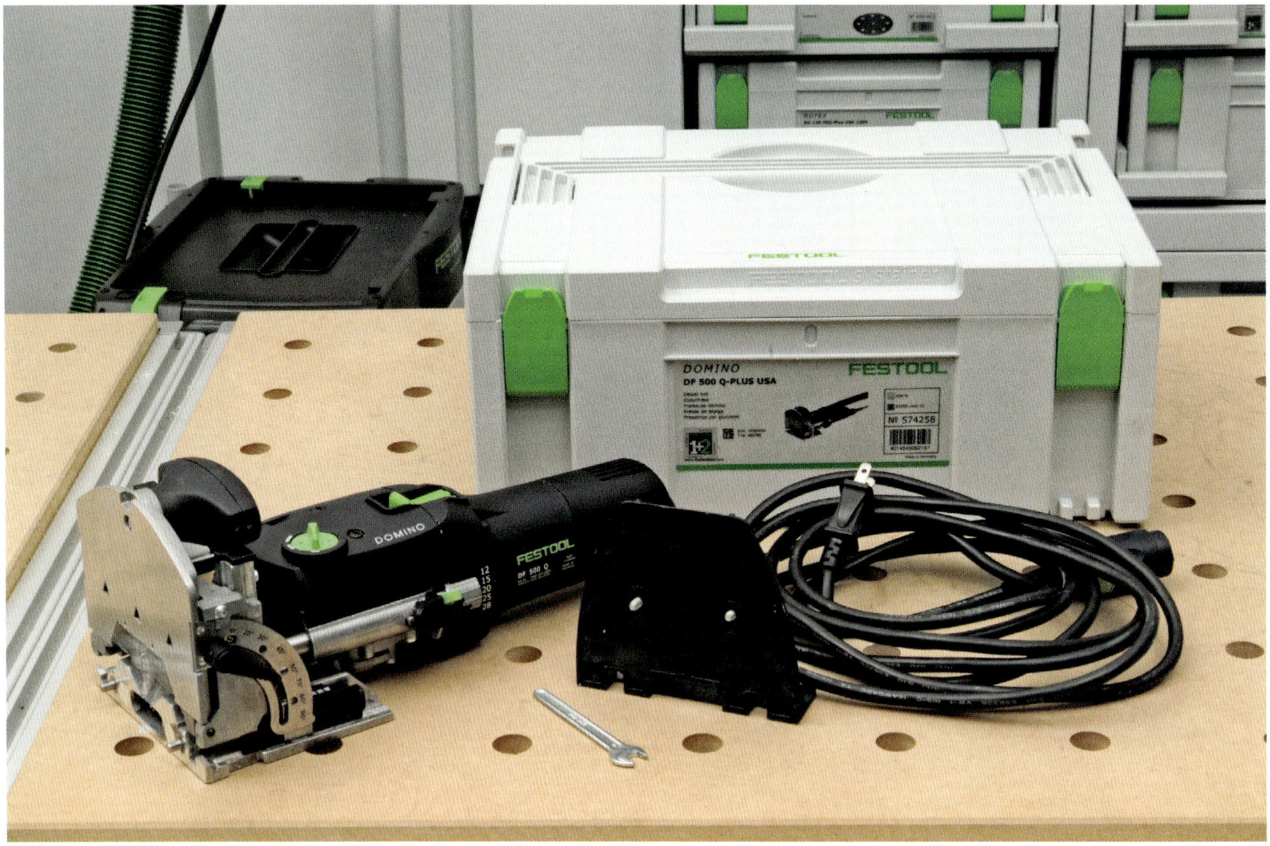

The basic Domino® System package comes with the Domino®, the Systainer®, a power cord, a support bracket, and a wrench.

Domino® set includes a Trim Stop, on the left, and two Cross Stops, in addition to the items in the basic package. Their use will be explained later.

The key to the Domino® system is the Domino® itself. Made of solid beech, the domino-shaped pieces come in five sizes: 5mm x 30mm, 6mm x 40mm, 8mm x 40mm, 8mm x 50mm, and 10mm x 50mm.

The Domino® tenons are available in a Systainer®, containing the complete assortment of 1105 beech Domino® tenons, plus the four carbide-tipped cutters necessary for making the Domino® mortises, 5mm, 6mm, 8mm, & 10mm.

The Domino® tenon offers several advantages over the more traditional methods. The common biscuit, left, lacks strength and, being a pressed wood product does not have the structural integrity of the more substantial, solid Domino® tenon. While the dowel is solid wood, it is harder to fit accurately and tends to rotate in position, while the Domino® system provides extreme accuracy and the tenon's shape allows it to stay in place. In addition the shape of the Domino® tenon provides a larger gluing surface, and for outdoor use all sizes are available in sipo mahogany, a weather resistant wood, seen here at the right.

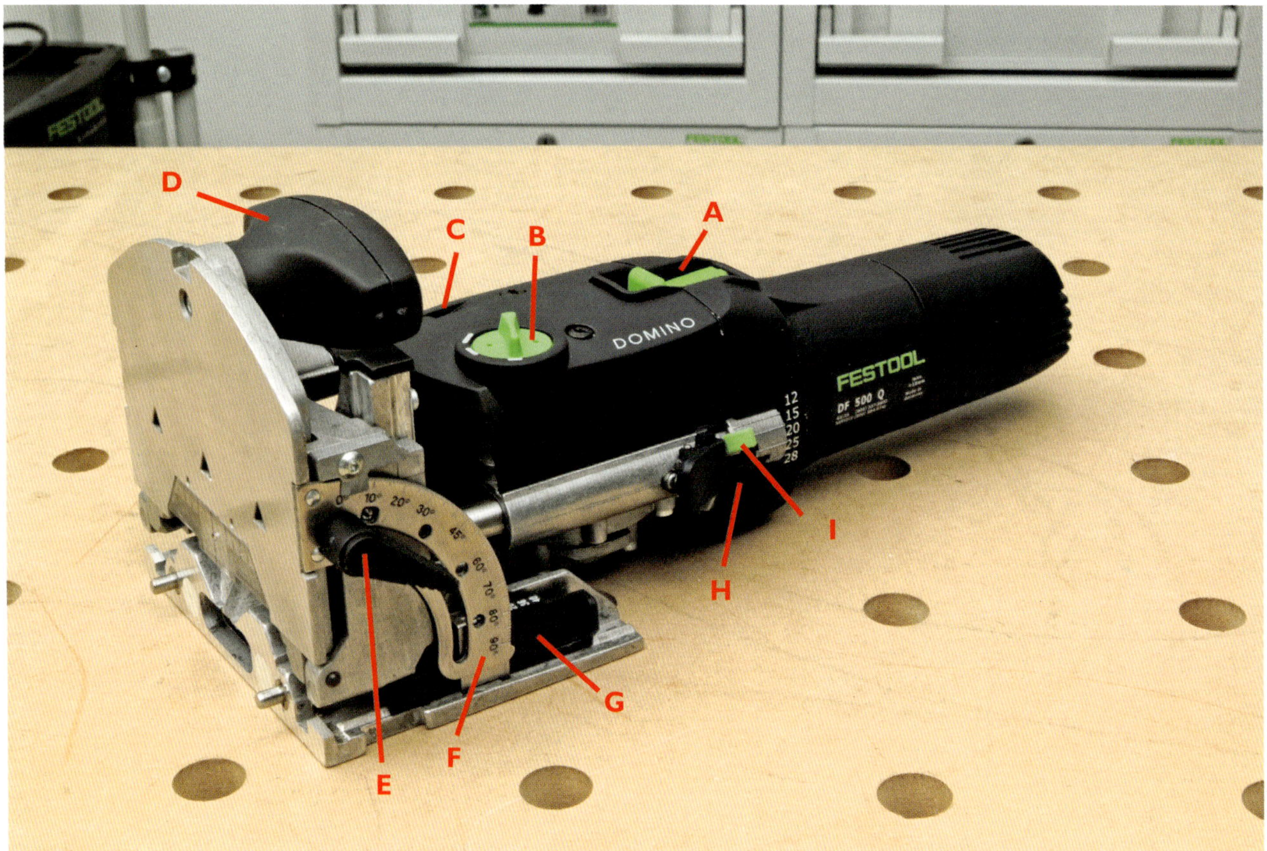

The machine overview.

A • ON/OFF switch

B • Rotary switch for setting the lateral tolerance of the Domino® mortise

C • Locking device for mounting the guide frame to the body.

D • Auxilary handle

E • Lock for the bevel gauge

F • Bevel gauge

G • Stepped gauge for setting material thickness

H • Plunge depth gauge

I • Plunge depth gauge lock lever

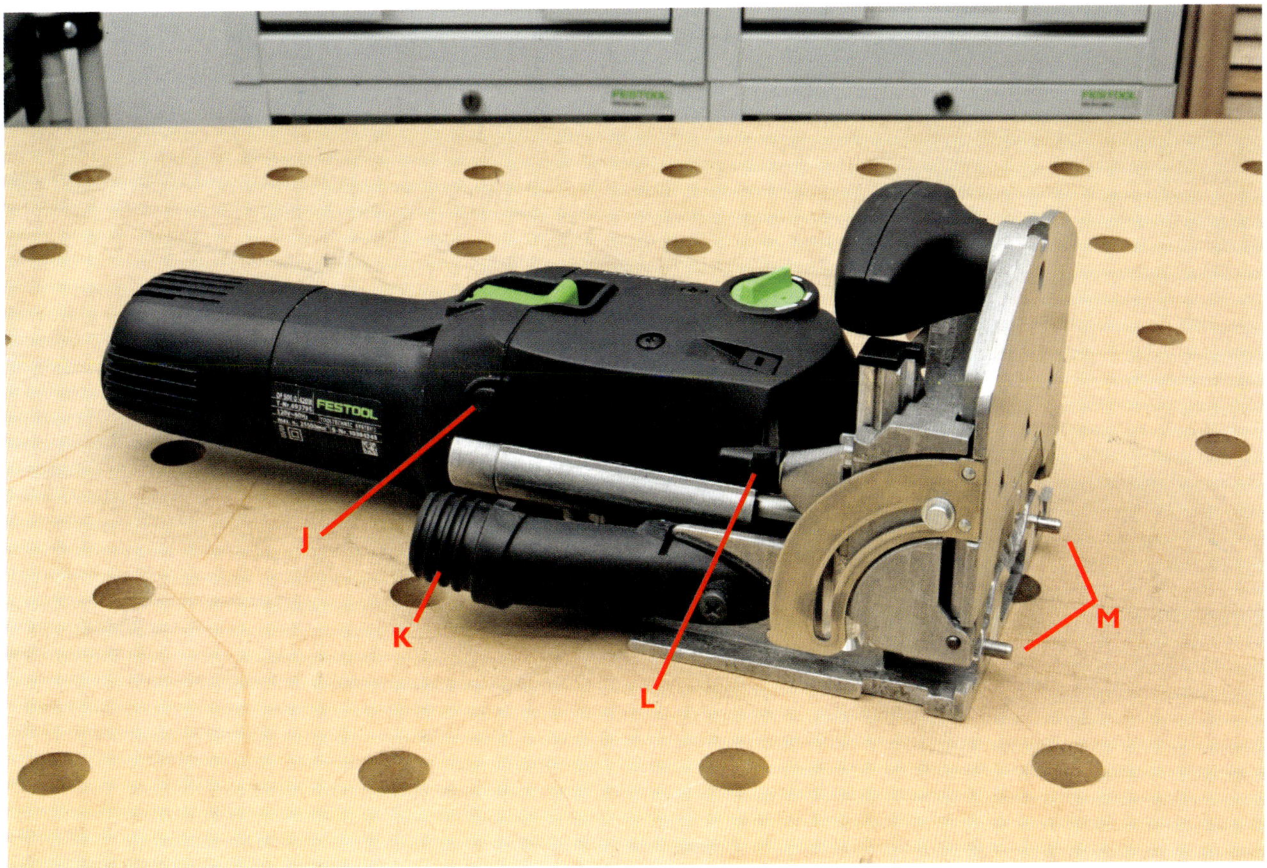

J • Arbor lock

K • Extractor hose nozzle

L • Clamping lever for joint height adjustment

M • Stop pins

The rotary switch on the top of the machine adjusts the tolerance of the mortise to be cut. The smallest mark denotes the tightest tolerance, producing a mortise with no lateral adjustability

The middle tolerance provides 6mm of lateral adjustability.

The widest mark produces the widest mortise, giving 10mm of lateral adjustability.

The locking on/off switch is also on the top.

The depth gauge is on the side. Set it by depressing the black lock lever and moving the button to the appropriate depth.

The depth of the mortise depends on the particular project. This 30mm Domino® tenon is to be inserted equally into each board. The depth in this case will be set a 15mm; 15mm + 15mm = 30mm.

At this corner a stronger 40mm Domino® tenon will give more structural integrity, but a 20mm mortise would go through the face board. Instead make a 12mm mortise in the face board and a 28mm mortise into the edge of the abutting board. 12mm + 28mm = 40mm.

The tool is now set at 12mm. With a little mathematics, the combination of these depth settings will be adequate for any situation.

The dust extraction hose attaches here.

The power cord attaches at the back. It is pushed in and given a quarter-turn to lock it in place.

To access the cutter, use the wrench to lift up the tab at the top. The body will be released from the fencing system.

Depress the arbor lock on the side of the body.

Use the wrench, turn counterclockwise to loosen the cutter.

Unscrew it by hand.

Replace with a new cutter. Hand screw until snug...

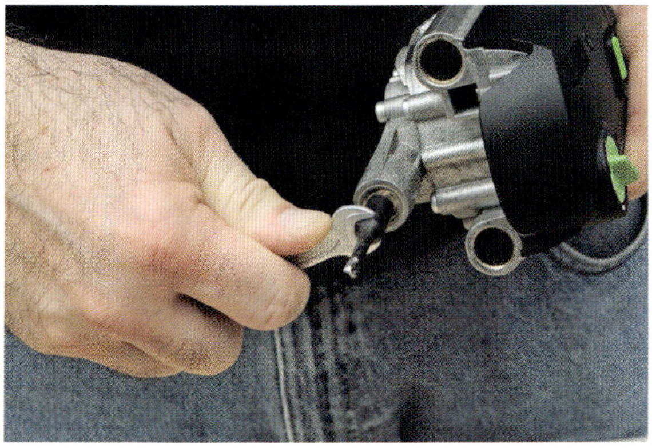

...and tighten it with a wrench. Use moderate pressure. You do not need to over-torque it.

This allows you to engage the longer rod first, and then the shorter one, eliminating a lot of trial and error. When the rods are in place the fence should easily snap back into place.

The guiding rods in the fence are precision ground for accurate operation. To enable easier reassembly, one is longer than the other.

The fence. Its features are easier to explain when it is free from the motor body.

The stepped gauge on the side is set to the thickness of the wood, which automatically sets the cutter at the center of the thickness.

In this case it is set for 20mm, the setting for most 3/4" wood. The number shows in the window.

On the back of the fence is a second scale that measures the distance from the face plate of the the fence to the center of the cutter, which references from the top of the wood. With the stepped scale set at 20mm, this scale automatically reads 10mm.

When the depth of the cut is established, lock it into place with the lever on the opposite side.

Custom settings are also easy to achieve using the scale on the back of the fence. First, back off the stepped scale so no number shows in the window.

Loosen the lock and adjust the plate to the desired depth. The number on the scale indicates the distance from the plate to the center of the cutter.

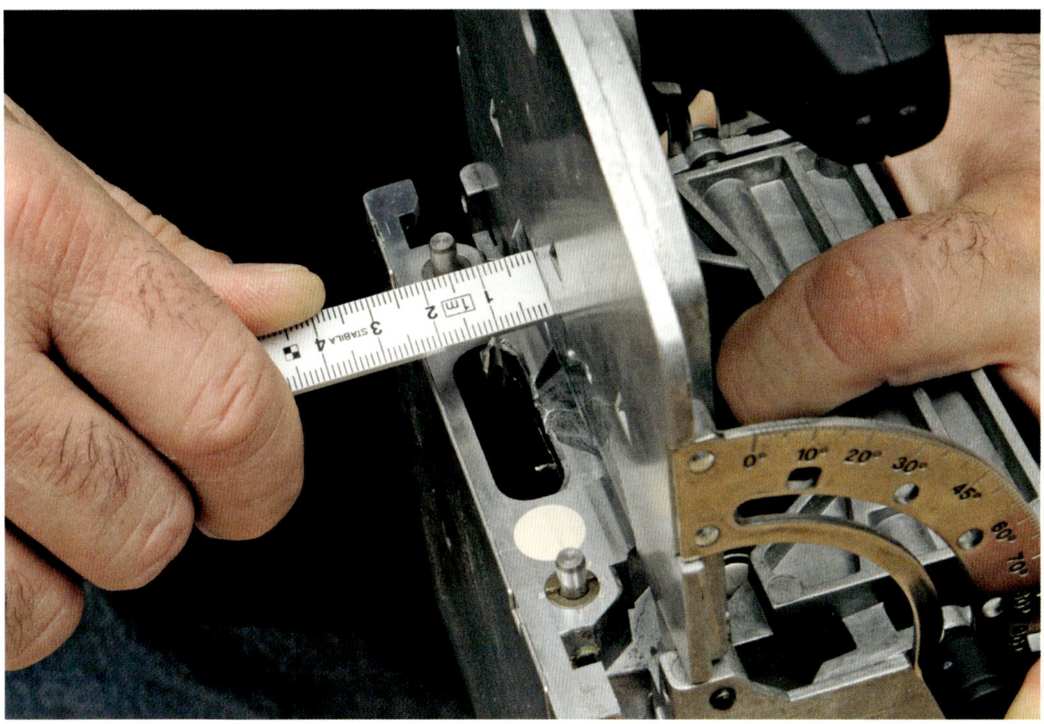

It is an accurate and simple way to set the cutter for custom work.

To place the mortise in the middle of this 36mm piece of lumber, simply use the step scale. To offset it 12mm from the surface, use one of two easy methods for setting the Domino®. The first is the mathematical. Simply the measure the distance to the center of the mortise, marked here at 12mm.

Then loosen the lock and set the scale to 12mm. The center of the mortise will now be 12mm from the surface.

The second method is the "eyeball" way. With the line drawn at the desire depth, simply place the plate on the surface of the wood, loosen the lock, and align the center of the pin with the line. These adjustments were made with the fence removed from the tool body. It is not necessary, but it is lighter and easier to maneuver.

The plate can be adjusted to any angle and has stops at 0, 22.5, 45, 67.5, and 90 degrees. Unlock and move to the required position.

For mitre-fold joinery, the stepped gauge should set with a black window the stepped gauge should be set all the way back until it can go no further. This setting can be confirmed with the top sight window showing no digits. For 3/4" stock the plate should always be zeroed out in its lowest position.

Other thicknesses require custom settings. Hold the Domino®
in position on the edge of the joint.

Mark the center of the Domino®.

Hold the wood to the fence and adjust so the pin is centered on the mark.

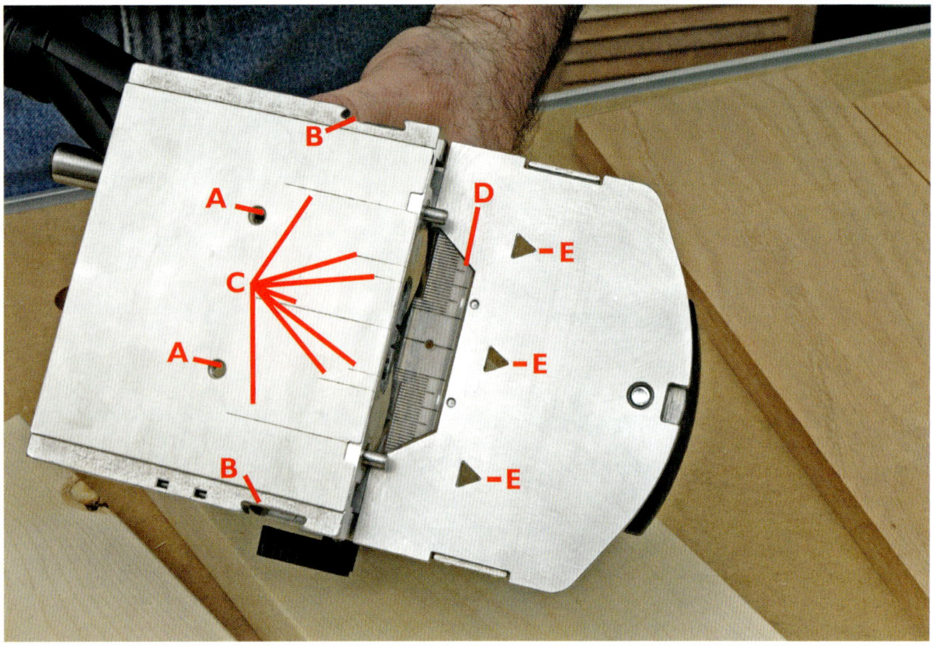

The bottom of the plate has several features.

A • At left are threaded holes for attaching the support bracket.

B • At the edge are slots for the attaching the cross stop

C • Scribe lines are etched in the back plate used for referencing the bit and pins

D • Two spring-loaded stop pins

E • A gridded plastic lens for referencing witness marks.

F • Three cutouts referencing the center of the cutter and the inside edges of the two pins

The top of the plate has two slots for attaching the trim stops and the handrail fence. Cast into the housing above the gridded lens is the center line of the cutter.

The pins provide an accurate yet simple referencing method for multiple mortise operations.

Example: The Basic Joint

The process begins by marking two abutting corners on the top surface of the boards that will be joined. It is the only mark you will need to make.

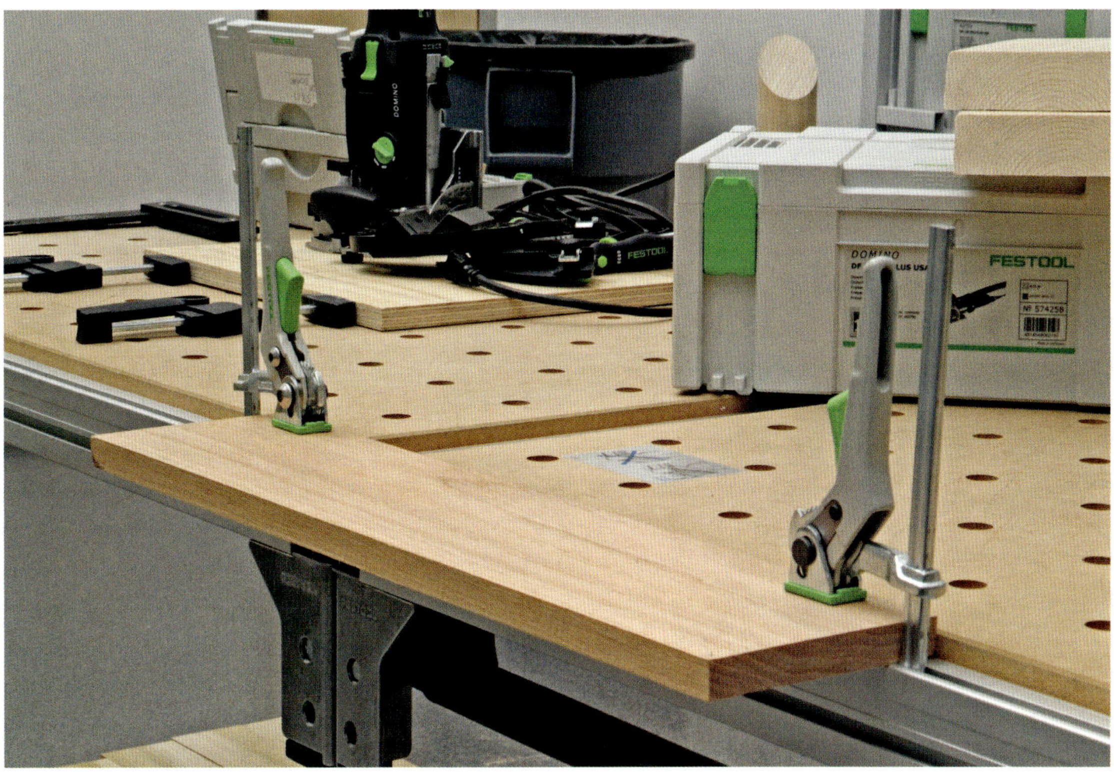

To work off the face of the material, clamp it to the table so it overhangs the edge.

Example: The Basic Joint

The first cut is an indexing cut, so set the tolerance to tightest level.

When using a 5mm x 30mm Domino® tenon, the depth gauge is set to 15mm.

The first cut is made with the pin shouldered against the end of the work.

Example: The Basic Joint

With the alignment set, plunge the cutter slowly into the wood. Moving at 24,000 rpm and oscillating back and forth, the Domino® does the rest of the work.

The first mortise is cut. This will act as a reference for the remaining cuts.

Example: The Basic Joint

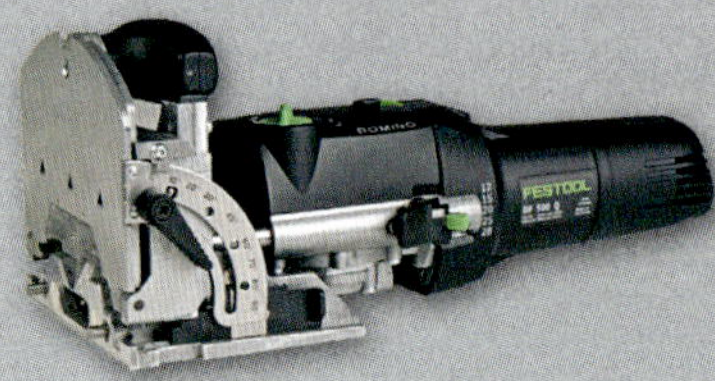

Align the pin so it shoulders on the inside of the first mortise.

Now move the rotary switch to the medium setting.

Example: The Basic Joint

Make the second cut.

Continue in the same way down the edge of the board.

The result is equidistance precise mortises. Of course, there are far more than will be needed for an actual joint. More on that later.

Now for the other board. Reset the tolerance to its tightest.

Locate the marked end, showing the top surface.

Example: The Basic Joint

This time the work will move in the opposite direction, left to right. Shoulder the left pin on the end of the board and make the first mortise.

Remember to change the tolerance to the medium level for the remaining mortises, as on the first board. Continue down the edge of the board.

Example: The Basic Joint

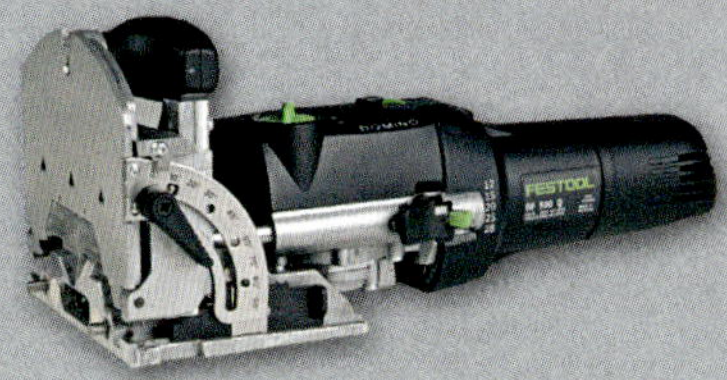

The result is perfectly aligned mortises ready for the Domino® tenons.

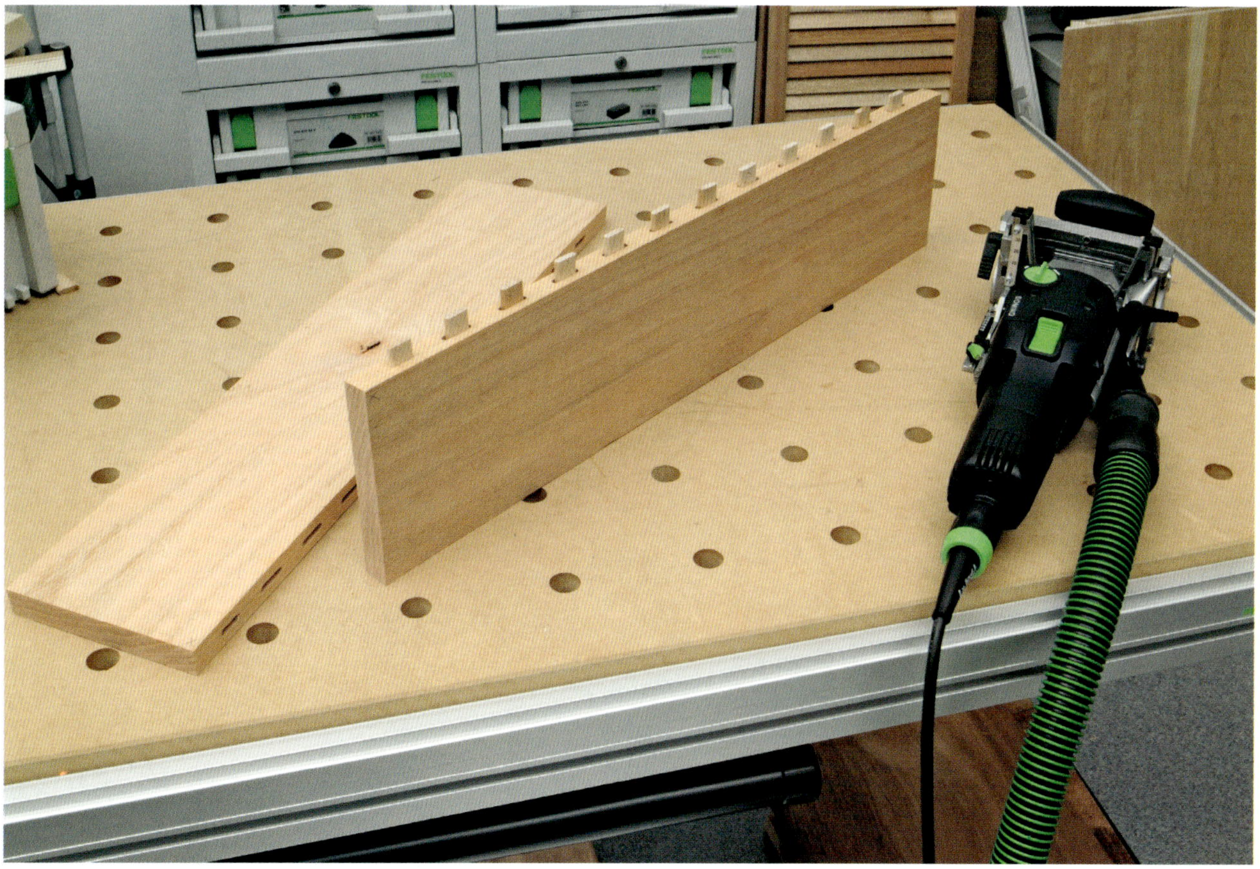

Place the Domino® tenons in one edge.

Example: The Basic Joint

Join the boards together beginning with at the marked end, where the reference tenon is. The others, being cut with more tolerance, should move easily into place.

The result is an almost invisible, very strong joint.

The Cross Stop

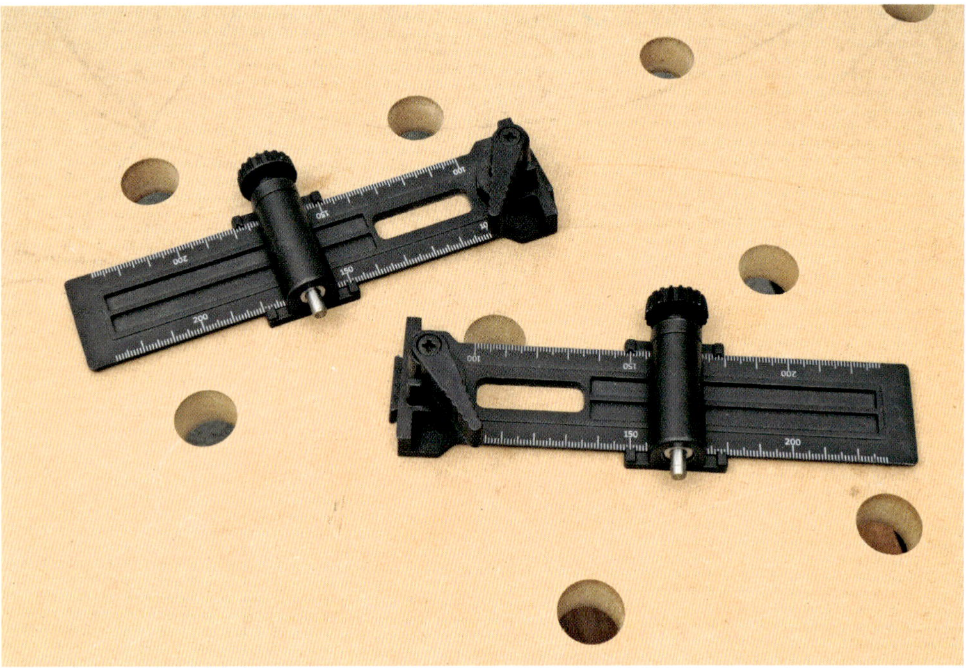

To set more space between mortises, the Cross Stop has been developed. It attaches to the fence, and has adjustable pins that can be set as needed.

Insert the bracket into the notch on the bottom of the plate…

...and lock in place.

The Cross Stop

The pins are on sliding brackets. Loosen the knob and slide to the desire location. The distance from the center of the cutter to the center of the pin appears on the ruler in the notch on the machine side of the pin. Here it is set at 150 mm. Set both pins at the same distance.

With a Cross Stop attached to both sides, we now have four indexing pins.

The Cross Stop

As before the first mortise is made using the pin on the plate. Shoulder it to the wood, and with the cutter set to its closest tolerance make the index mortise.

Adjust the tolerance to the medium setting, place the pin of the Cross Stop in the index mortise and make the second mortise.

Continue down the board

Repeat on the other board, starting at the marked end with an index mortise, and continuing using the Cross Stop pins and the medium tolerance.

The result.

Insert the Domino® tenons and begin joining at the index mortise.

Work your way up…

…to a perfect joint.

The Trim Stop

The Trim Stop allows you to work with narrow material, 7/8" to 2-3/4" wide, 22 to 70 mm.

A groove in the Trim Stop fits over the nose of the front plate…

The Trim Stop

...allowing it to slide into place.

Brackets on the Trim Stop fit into slots on the top surface of the front plate and are secured with the knobs.

The Trim Stop

The bottom of the Trim Stop consist of two independently sliding guides that create a channel between them. The green lock-down knobs hold them in place. The other screws are guide screws and are factory-set to allow free movement of the plates.

To set the Trim Stop, begin by marking the center of the board to be joined.

The Trim Stop

Loosen the sliding guides.

Center the wood in the guide by aligning the center line with the scribed center line of the Domino® plate.

Next, set the thickness gauge to the desired level. In this case one tenon is being used for a butt joint and needs to be centered.

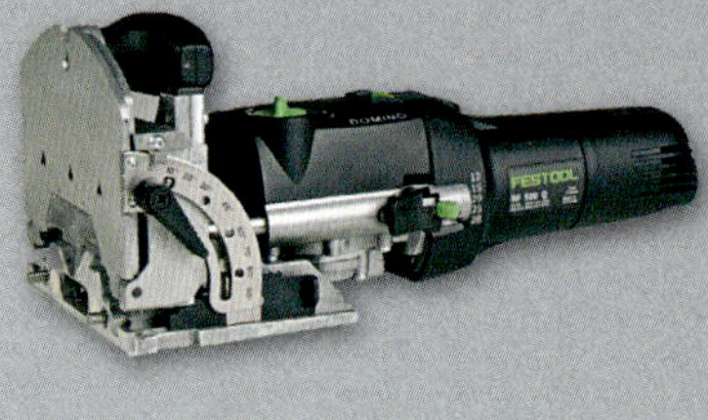

Example: The Basic Butt Joint

The wood is 48mm thick. The step gauge only goes to a thickness of 40mm, so I'll use the gauge on the fence and set it at 24mm. Important: this depth will stay the same throughout the operation, even after the Trim Stop is removed for the cut on the other board.

This joint uses a 10mm x 50mm Domino® tenon, so requires a change to a 10mm cutter with the plunge depth set to 25mm, half the tenon length.

Example: The Basic Butt Joint

One side of the joint should be set to at the tightest tolerance. Since there is less movement in the end grain, it is a reasonable choice for this. Set the tolerance, align the tool in the Trim Stop and make the mortise.

The result is a centered, strong tenon.

Example: The Basic Butt Joint

The tenon placement in the cross piece involves aligning the witness marks with the tool. There are three reference points for the center of the mortise: the center cut-out triangle; the scribed lens, and the cast reference on the fence. Any one will give you the proper alignment, but you may wish to double check yourself with another.

With the tolerance set at medium and the tool aligned, make the mortise.

Example: The Basic Butt Joint

Make sure your face references match and join the pieces.

Align the witness lines.

The result is a perfectly square joint. It is strong because of the solid wood construction of the Domino® tenon, and will not turn because of its unique shape.

The Handrail Fence

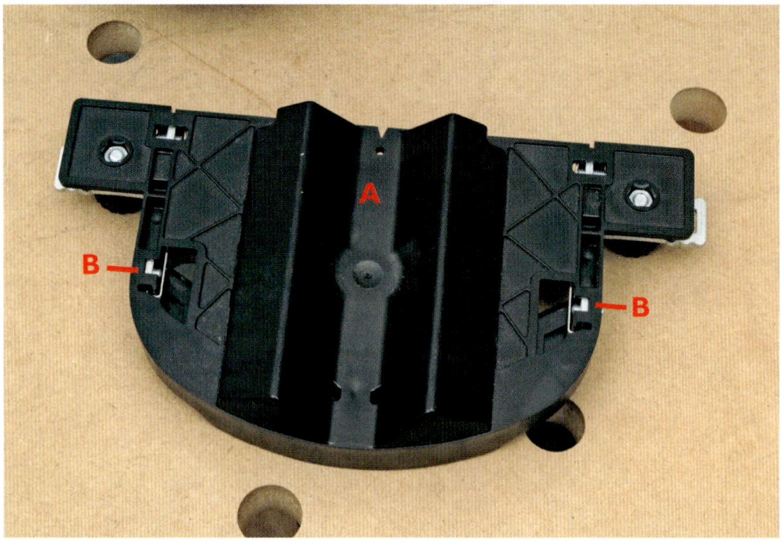

The Handrail Fence accommodates round pieces between 35mm (1-3/8") and 60mm (2-3/8"). The central depression in the underside (A) holds the material. The screws nearest the camera (B) on each side are for adjustment in particular alignment situations.

The fence slides over the nose of the plate and clamps in place, like the Trim Stop did.

Example: A Right Angle Handrail Joint

For demonstration, these two pieces of 50mm railing will be joined at a right angle. Each has been cut at 45 degrees…

,,,to go together like this.

With the pieces in place, hold the tenon in the desired position

Example: A Right Angle Handrail Joint

Mark the center of the tenon. Only one side needs to marked as the tool will be set to it and will automatically duplicate the cut on the other piece.

Set the bevel gauge to match the angle of cut, in this case 45 degrees.

Example: A Right Angle Handrail Joint

Holding the piece in position in the Handrail Fence, so it comes against the face fence, set the thickness of the cut by adjusting the fence until the center of the pin aligns with the mark.

Lock the fence in place.

Example: A Right Angle Handrail Joint

Replace the fence on the machine. The Domino® tenon is 10mm x 50mm so the plunge depth should be set at 25mm, allowing equal lengths of tenon for each piece.

The rotary switch should be set for no extra lateral movement, the narrowest setting. In this situation there will be a single pin joint, with no room for error, so the narrowest setting should be used on both pieces.

Example: A Right Angle Handrail Joint

When the piece is held snuggly against the face, and resting in the Handrail Fence, it will automatically assume the correct position. Any turn will create a gap between the piece and the face which is readily observable.

The result.

Example: A Right Angle Handrail Joint

Repeat the cut on the other piece.

Insert the tenon…

...and join the two pieces. This creates a solid and, for a hand rail, safe joint with good structural integrity.

The Support Bracket

Often the Domino® is used in the vertical mode. When used this way, there is a tendency to tip one way…

…or another, as can be seen clearly in the edge position.

The Support Bracket

For this reason, a Support Bracket is included with every Domino® machine. It screws into the bottom of base plate.

This gives much greater stability for the vertical cutting position.

Example: A Butting Case Joint

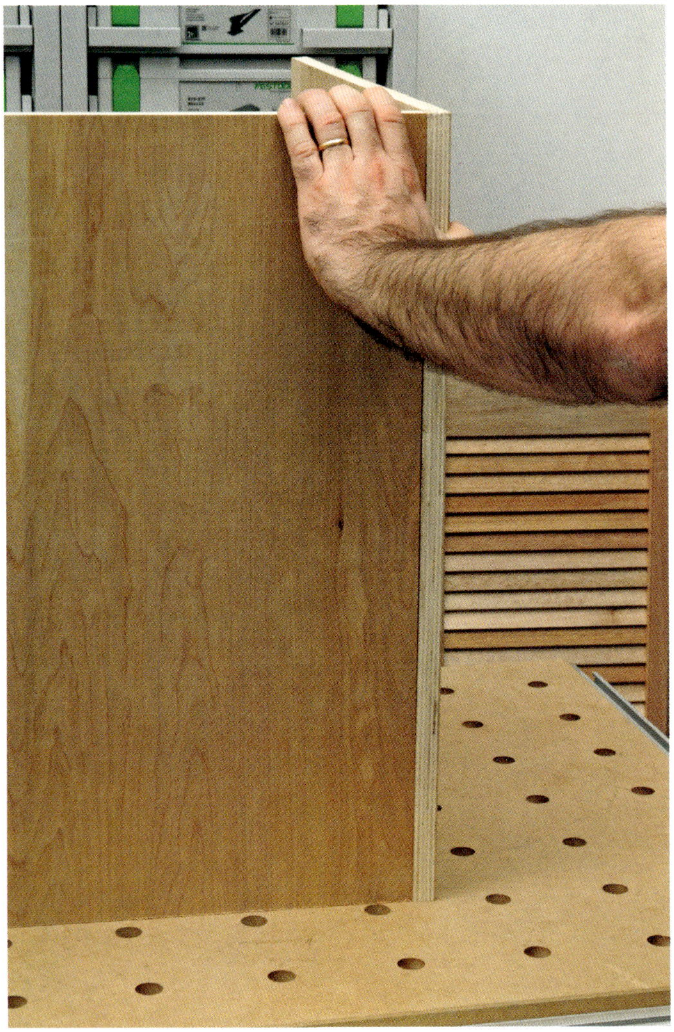

One place this is useful is a butting case joint.

With the corners aligned, mark an X on the facing side. Mortises will be cut into the edge of this piece.

Mark a Y on the adjoining side, where mortises will be cut into the face.

Example: A Butting Case Joint

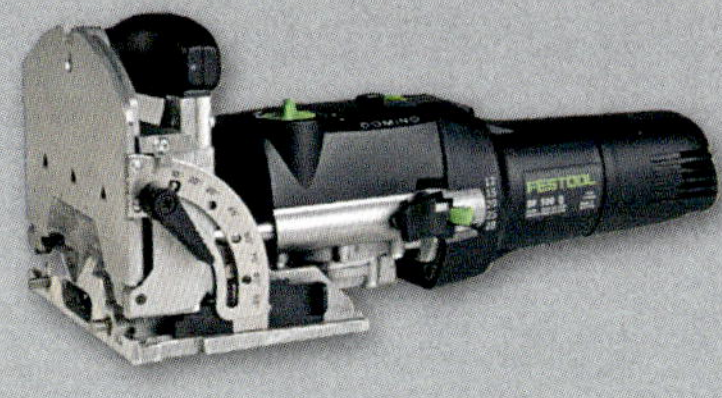

For a 5mm x 30mm Domino® tenon set the plunge depth to 15mm.

The thickness of the wood is 20mm, which can be set with the step gauge.

Example: A Butting Case Joint

The narrowest tolerance is set for the first mortise on each edge, then reset to medium tolerance for the rest.

The mortises will be spaced 4" on center. To achieve this use the Cross Stops, setting each pin to 100mm.

Starting with the edge mortises on the piece marked X, shoulder the machine pin and make the first, tight tolerance, mortise.

Adjust the tolerance to medium and use the Cross Stop pin to space the remaining edge mortises.

Example: A Butting Case Joint

Continue down the edge.

The face mortises are done similarly. Change the tolerance to fine for the first mortise, shouldering the machine pin for position.

Example: A Butting Case Joint

Change the tolerance to medium and continue with the mortises down the edge, using the pin in the Cross Stop for position.

The result.

Place the Domino® tenons and join the sides.

Example: A Cabinet Corner

The beveled cabinet corner begins by cutting the pieces at a 45 degree angle. Lay the two pieces to be joined side by side and mark with Xes in the adjoining top corners.

Set the bevel gauge angle to 45 degrees.

Example: A Cabinet Corner

Defeat the stepped automatic height gauge by backing it off. Black should appear in the window.

Set the custom height gauge to zero.

Example: A Cabinet Corner

The Domino® tenon is 5mm x 30mm, so the plunge depth is set at 15mm.

As on the last example, the mortises will be 4" on center, so the pins on the Cross Stops are set to 100mm.

Example: A Cabinet Corner

With the tolerance at the finest setting start at the X-ed end and make the first mortise, shouldering the machine pin on the edge of the piece.

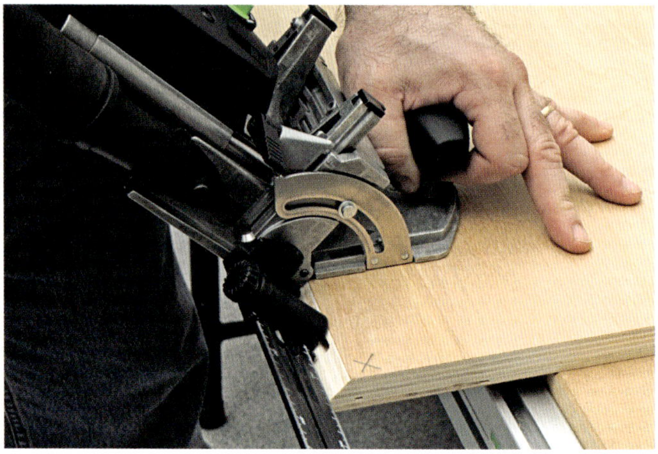

Change the tolerance to medium and shoulder the Cross Stop pin in the first mortise. Continue this way down the edge.

The result

Repeat the process from the other piece, again starting at the X-ed end. Again, the first, index mortise should be done at fine tolerance.

Example: A Cabinet Corner

Insert the tenons.

Join the two pieces.

The joint complete

Example: A Compound Miter Joint

A compound miter joint is made easy with the Domino® system, giving a solid joint with no play in it. For the demonstration this is a 45 degree angle with a 22.5 degree bevel.

Mark two adjoining corners with Xes.

Subtract the bevel angle (22.5) from 90 degrees to arrive at the angle of the fence, 67.5 degrees.

Example: A Compound Miter Joint

The Domino® tenon is 5mm x 30mm so the plunge depth is set a 15 and the cutter is 5mm.

The stock is 3/4 inches, so the automatic step thickness gauge is backed-off with black appearing in the window, and the fence gauge is zeroed out.

Example: A Compound Miter Joint

For this all the mortises will be cut at the tightest tolerance.

Shoulder the pin and cut the first mortise, starting at the X.

Example: A Compound Miter Joint

Move the pin to the first mortise, shoulder it and continue down the edge.

The result.

Example: A Compound Miter Joint

Repeat on the facing edge, again starting at the X-ed end.

Insert the Domino® tenons, align,...

...and push together.

The result.

Adjusting the Pins

Even the best tools show wear over time. This is true even of the Domino®. In time the pin system may need adjustment. The symptom of this is a joint that misaligns, as seen here.

The front left pin is on eccentric cam so that it can be adjusted when necessary.

Adjusting the Pins

The locking screw for the pin is accessed from the side, just above the foot plate.

Loosen the screw with a 2.0mm hex key.

Adjusting the Pins

Depress the pin with the screw driver.

Adjusting the Pins

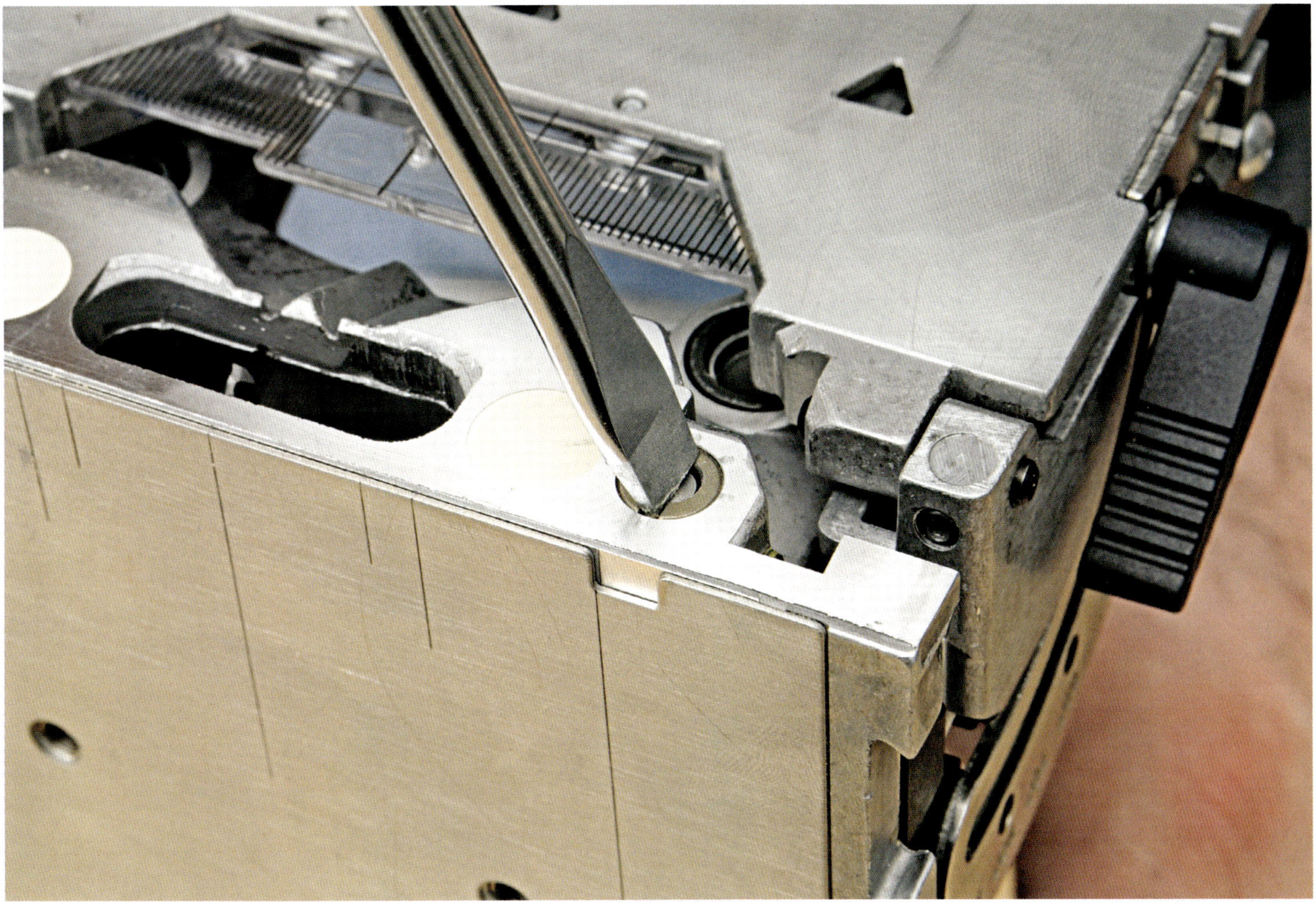

The collar around the pin is slotted and the surface around it is indexed. At the factory the slot aligns vertically. The adjustment will require some trial and error. Turn the collar one notch in either direction and test the placement using two pieces of scrap wood. If the alignment is worse, turn the collar the other direction and test again. If is better, but a little too much or too little, tweak it as necessary. When the alignment is perfect, tighten the lock screw.

The Dust Extraction System

Dust control is an integral part of the Festool system and design philosophy. It offers a full line of dust extractors designed to for the workshop or jobsite, to work with the Festool tools to optimize their operation and effectiveness at the same time as they provide a safe, clean work place.

The cleaning sets come in metal or plastic and the choice of anti-static or non anti-static hoses in three diameters to meet all your tool and dust extraction requirements. A wide assortment of connectors, reducers, and special sleeves allow you to tailor your system.

Festool offers a wide selection of filtration options for ultimate flexibility. In addition to the standard disposable filter bags, Festool offers a longlife, re-usable bag. There are four main filters to choose from; standard, long-lasting, wet and HEPA (only available on CT 22 E and CT 33 E). An optional dirt trap is available for wet applications.

Sys-Docks and optional hose hooks, holders, and garages (for the CT 22 E and CT 33 E) are designed for efficient storage of cleaning accessories and Festool Systainers. The flat tops offer added space for tools while you work.

With the optional boom arm, the hoses and cords are kept up and out of the way, opening up the work space.

The CT 22 E and CT 33 E can be equipped for connecting to pneumatic tools. Compressed air service units, auto-start conversion kits and IAS2 single and double adapters make the CT even more versatile.

Features

· **Tool-triggered activation** for automatic dust extraction.

· **Flat tops with locking latches** to attach Systainers, for ease of transport and storage.

· **Simple, pressure-fit hose connectors** for switching tools quickly.

· **Antistatic hoses are standard** to reduce dust build-up.

· **Ergonomically designed carrying handles and oversized wheels** for ease of movement around the jobsite.

· **Built-in cord storage** reduces tripping hazards.

· **HEPA filters** eliminate 99.99% of particulates from the air (Only available on CT 22 E and CT 33 E, where they are standard equipment)

· **The motor is completely separate from the dust bin,** significantly prolonging the life of the motor.

· **Adjustable cubic feet per minute (CFM)** setting for better results.

Specifications

Model name	CT MINI	CT MIDI	CT 22 E	CT 33 E
Item	**583 360**	**583 376**	**583 366**	**583 368**
Standard package includes				
Filters	I micron	I micron	HEPA	HEPA
Filter bag	2	2	I	I
Antistatic suction hose	I I/16" x I I.5' (27mm x 3.5m)	I I/16" x I I.5' (27mm x 3.5m)	I I/16" x I I.5' (27mm x 3.5m)	I I/16" x I I.5' (27mm x 3.5m)
Container/filter bag capacity	2.6 gal/2 gal 10 liters/7.5 liters	3.96 gal/3.3 gal 15 liters/12.5 liters	5.8 gal/5.3 gal 22 liters/20 liters	8.7 gal/7.9 gal 33 liters/30 liters
Specifications				
Depth x width x height	17.5" x 13.5" x 16.5" (440 x 340 x 420 mm)	17.5" x 13.5" x 18.5" (440 x 340 x 470mm)	24" x 15" x 17" (625 x 380 x 440 mm)	24" x 15" x 22" (625 x 380 x 550 mm)
Weight	21 lbs. (9.5 kg)	21.4 lbs. (9.7 kg)	26.5 lbs. (12 kg)	32 lbs. (14.5 kg)
Noise (sound level)	62 dB (low setting) 72 dB (maximum setting)	62 dB (low setting) 72 dB (max. setting)	62 dB (low setting) 72 dB (maximum setting)	62 dB (low setting) 72 dB (maximum setting)
Maximum suction capacity	99 CFM (2807 l/min)	99 CFM (2807 l/min)	134 CFM (3800 l/min)	134 CFM (3800 l/min)
Maximum vacuum	80" static water lift (200000 Pa)	80" static water lift (20000 Pa)	90" static water lift (23000 Pa)	90" static water lift (23000 Pa)
Power consumption	400-1200 watts / 3.3-10 amps 120 v AC	400-1200 watts / 3.3-10 amps 120 v AC	350-1200 watts / 2.9-10 amps 120 v AC	350-1200 watts / 2.9-10 amps 120 v AC

Left to right: CT Mini (99 cfm, 2 gal. cap.), CT Midi (99 cfm, 3.3 gal. cap.), CT 22 E (134 cfm, 5.3 gal cap.), and CT 33 E (134 cfm, 7.9 gal. cap.). The smaller two units are great for job sites, being easily portable and small enough to get in tight spots. The CT 33 E is mainly a workshop unit, with large capacity and power. The CT 22 E is an all purpose unit, with power and size that are good for the shop and worksite.

Description/Application	Item	Compatability			
		CT MINI	CT MIDI	CT 22 E	CT 33 E
Hoses—Antistatic					
Antistatic versions are only effective when used with antistatic design (Festool's CT series) dust extractors.					
With rotating connector and reducing sleeve.					
1 1/16" x 11.5' (27 mm x 3.5 m) AS	**452 878**	●	●	●	●
1 1/16" x 16.5' (27 mm x 5 m) AS	**452 880**	●	●	●	●
1 7/16" x 11.5' (36 mm x 3.5 m) AS	**452 882**			●	●
1 7/16" x 16.5' (36 mm x 5 m) AS	**452 884**			●	●
1 7/16" x 23' (36 mm x 7 m) AS	**452 886**			●	●
1 15/16" x 13' (50 mm x 4 m) AS	**452 888**			●	●
1 15/16" x 13' (50 mm x 4 m) AS	**452 890**			●	●
Hoses—Non antistatic					
For use with cleaning sets or non-antistatic dust extractors.					
With rotating connector and reducing sleeve.					
1 1/16" x 11.5' (27 mm x 3.5 m)	**452 877**	●	●	●	●
1 1/16" x 16.5' (27 mm x 5 m)	**452 879**	●	●	●	●
1 15/16" x 11.5' (36 mm x 3.5 m)	**452 881**			●	●
1 15/16" x 16.5' (36 mm x 5 m)	**452 883**			●	●
1 15/16" x 23' (36 mm x 7 m)	**452 885**			●	●
1 15/16" x 8.25' (50 mm x 2.5 m)	**452 887**			●	●
1 15/16" x 13' (50 mm x 4 m)	**452 889**			●	●
Hose Sleeves—Rotating connector					
Hose connector to fit inlet of dust extractor, external dia. 2 9/32"					
Antistatic version for D 50 suction hose	**452 896**			●	●
Antistatic version for D 36 suction hose	**452 894**			●	●
Antistatic version for D 27 suction hose	**452 892**			●	●
Non-antistatic version for D 50 suction hose	**452 895**			●	●
Non-antistatic version for D 36 suction hose	**452 893**			●	●
Non-antistatic version for D 27 suction hose	**452 891**	●	●	●	●
Hose sleeves--Reducing sleeves					
Hose connector for tools or cleaning sets.					
With rotating adapter for D 27 suction hose *Replacement. Tool end of extraction hose for our standard 27mm hose.*	**487 071**	●	●	●	●
Antistatic version with rotating adapter for D 36 suction hose *Replacement. Tool end of extraction hose for mid-sized 36mm hose.*	**487 721**			●	●

Accessories

Description/Application	Item	CT MINI	CT MIDI	CT 22 E	CT 33 E
Non-antistatic hose reducer	**452 897**			●	●
For reducing from 50mm to 36mm hose or when connecting 36mm to 27mm hose.					
Antistatic angle adapter	**456 806**	●	●		
Elbow hose end for connection in CT MIDI or CT MINI					
Antistatic connector sleeve	**493 047**			●	●
To connect suction hoses D 50/D 36/D 27.					
Antistatic Y-piece with blanking plug	**452 898**			●	●
To connect two suction hoses to the dust extractor.					
Blanking plug	**452 899**			●	●
For retrofitting on CT 23/33, to seal off inlet during transportation.					
Tubes—Curved					
Stainless steel	**452 900**	●	●	●	●
To connect tubes to extraction hose for comfortable grip.					
Polypropylene (PP) with air regulating slide to adjust suction power	**452 901**	●	●	●	●
To connect tubes to extraction hose for comfortable grip.					
Tubes—Extension					
Stainless steel	**452 902**	●	●	●	●
3 piece, 37 1/2" (950 mm) D 36					
Non-antistatic polypropylene (PP)	**452 903**	●	●	●	●
3 piece, 37 1/2" (L= 970 mm) D 36					
One-piece anodized aluminum	**447 599**			●	●
43 1/2" (L= 1100 mm) D 50					
Non-antistatic plastic	**440 412**			●	●
2-piece, 38" (L= 970 mm) D 50					
Floor nozzles/brushes					
Industrial floor nozzle	**452 908**	●	●	●	●
High-quality industrial design; aluminum. Width 14 5/8" (370 mm); 2 casters, 2 brush inserts.					
Large Industrial floor nozzle	**452 910**	●	●	●	●
High-quality industrial design; aluminum. Width 17 11/16" (450 mm); 2 casters, 2 brush inserts.					

Compatability: CT MINI, CT MIDI, CT 22 E, CT 33 E

Accessories

Description/Application	Item	Compatability			
		CT MINI	CT MIDI	CT 22 E	CT 33 E
Standard floor nozzle *Plastic. Width 11 13/16" (300 mm) 2 caster, 2 brush inserts*	**452 907**	●	●	●	●
Workshop floor nozzle *Plastic. Width 14 5/8" (370 mm) 2 caster, 2 brush inserts*	**452 909**	●	●	●	●
Interchangeable floor nozzle *4-piece, plastic. With 3 replacement inserts (brushes, rubber lips, carpet)*	**452 906**	●	●	●	●
Multi-purpose floor nozzle *With retractable brush inserts.*	**452 911**	●	●	●	●
Turbo suction brush *For carpet and hard surface with 4 casters. Width 10 5/8" (270 mm). Suction-activated rolling brushes, requires 36 mm hose.*	**450 644**			●	●
Special nozzles/brushes					
Suction brush *Plastic, dia. 2 13/16" (70 mm)*	**440 404**	●	●	●	●
Upholstery brush *Plastic, with brush insert width 4 11/16" (120 mm)*	**440 406**	●	●	●	●
Universal brush nozzle *Plastic, brush ring 4 11/16" x 1 5/8" (120 x 40 mm)*	**440 403**	●	●	●	●
Large crevice nozzle *Plastic, length 14 4/8" (370 mm)*	**452 912**			●	●
Crevice nozzle *Plastic, length 14 5/8" (370 mm)*	**452 904**	●	●	●	●
Bevel ended nozzle *Flexible rubber, length 7 1/2" (190 mm), outside diameter 1 5/8" (40 mm)*	**411 810**	●	●	●	●
Other accessories					
Spark trap *For use when operations my generate sparks, i.e. sanding or sawing metal.*	**484 733**			●	●
Handle *For CT 22 E & CT 33 E only. Improves mobility. Required for mounting Boom Arm 492 753.*	**452 921**			●	●
Suction hose holder *Raises hose to table height and prevents trip hazard.*	**487 072**			●	●
Hose hook *Hose minder for storage or to prevent hose from laying on the ground.*	**452 998**			●	●

Accessories

Description/Application	Item	Compatability CT MINI	CT MIDI	CT 22 E	CT 33 E
Dirt trap *Collection bucket facilitates disposing of liquids.*	452 925			●	
Dirt trap *Collection bucket facilitates disposing of liquids.*	452 926				●
CT adapter plug *Replacement. Adapts the 20 amp CT plug to fit in standard outlets.*	493 232	●	●	●	●
Boom arm *Keeps the cord and hose up and out of your way; requires handle (SB-CT 452 921) to be mounted onto CT 22/33 dust extractor.*	492 753			●	●
CT hose garage *Stores and protects hose and cord for transport.*	494 388			●	●
Boom Arm Tool Holder *Convenient storage and cradle of two electric sanders. For use with Boom Arm only. Includes wire rack and clamping knob.*	493 558			●	●
Filter Elements					
Main filter element *Standard 1 micron filter replacement-paper. 2 pieces.*	452 923			●	●
HEPA filter element *The highest standard for filtration. Filters particles to .3 microns.*	493 334			●	●
Wet filter element *Prevents foam buildup associated with wet extraction. 2 pieces.*	452 924			●	●
Wet filter element *Prevents foam buildup associated with wet extraction. 1 piece.*	454 805	●	●		
Longlife filter element *Cloth filter with special "M" fold prevents build-up. 2 pieces.*	454 869			●	●
Filter bags					
Filter bag *Paper filter bag captures dust. 5 pieces.*	456 772	●			
Filter bag *Paper filter bag captures dust. 5 pieces.*	494 105		●		
Filter bag *Paper filter bag captures dust. Includes dust cap for clean disposal. 5 pieces.*	452 970			●	

Description/Application	Item	CT MINI	CT MIDI	CT 22 E	CT 33 E
		Compatability			
Filter bag *Paper filter bag captures dust. Includes dust cap for clean disposal. 5 pieces.*	**452 971**				●
Longlife filter bag (re-usable bag) *Economical solution for capturing and eliminating large chips produced by routing or planing. (at least 500 fillings) Re-sealable.*	**456 737**			●	
Longlife filter bag (re-usable bag) *Economical solution for capturing and eliminating large chips produced by routing or planing. (at least 500 fillings) Re-sealable.*	**456 738**				●

The CT MINI & MIDI

The **CT MINI** and **CT MIDI** are Festool's most portable mobile dust extractors. Compact and lightweight, they have enough capacity for common jobsite tasks. The CT MIDI, somewhat larger in size, offers increased capacity for heavier on-site use. Both units have large wheels and double casters, that can easily negotiate obstacles in the shop or on the worksite. Generating 99 CFM of suction, and only 62 dB (on low setting), these CTs can quickly and quitely remove large amounts of debris. One disposable filter comes standard on the CT MINI and CT MIDI, with the option of additional dry and wet filters to meet your every need. The integrated hose garage keeps hoses and cords well organized and makes Systainer transport simple and efficient.

CT Mini & CT Midi Features:

• **Integrated Sys-Dock and hose garage.** Cords and hoses conveniently store in the onboard hose garage, while the flat top and locking latches allow you to attach one or more Systainers.

• **Filter bag.** Disposable, two-ply filter bags are particularly tear-resistant; an integrated dust cap prevents dust from spilling during disposal.

• **Tool-less main filter and automatic water level detector.** The main filter can be changed without using tools. The internal sensors monitor the fluid level, for wet applications, and switch the mobile dust extractor off automatically when the container is full.

• **Automatic tool start/Adjustable suction.** Smooth start, infinitely adjustable electronic air flow control and tool-triggered or manual on-off switch ensures convenient and efficient dust control.

• **Convenient carry handle.** The ergonomic handle makes maneuvering these small and light dust extractors a breeze.

• **Foot brake.** Quick on/off foot actuated brake prevents roll-off during usage or during transport.

• **Anti-static hose.** A 1" x 11.5' (27mm x 3.5m) antistatic hose comes standard.

• **Large wheels.** Large wheels, low center of gravity, and ergonomically balanced design allows you to move the CT effortlessly around the jobsite.

• **Mobility.** The manageable size makes it easy to transport or carry up stairs.

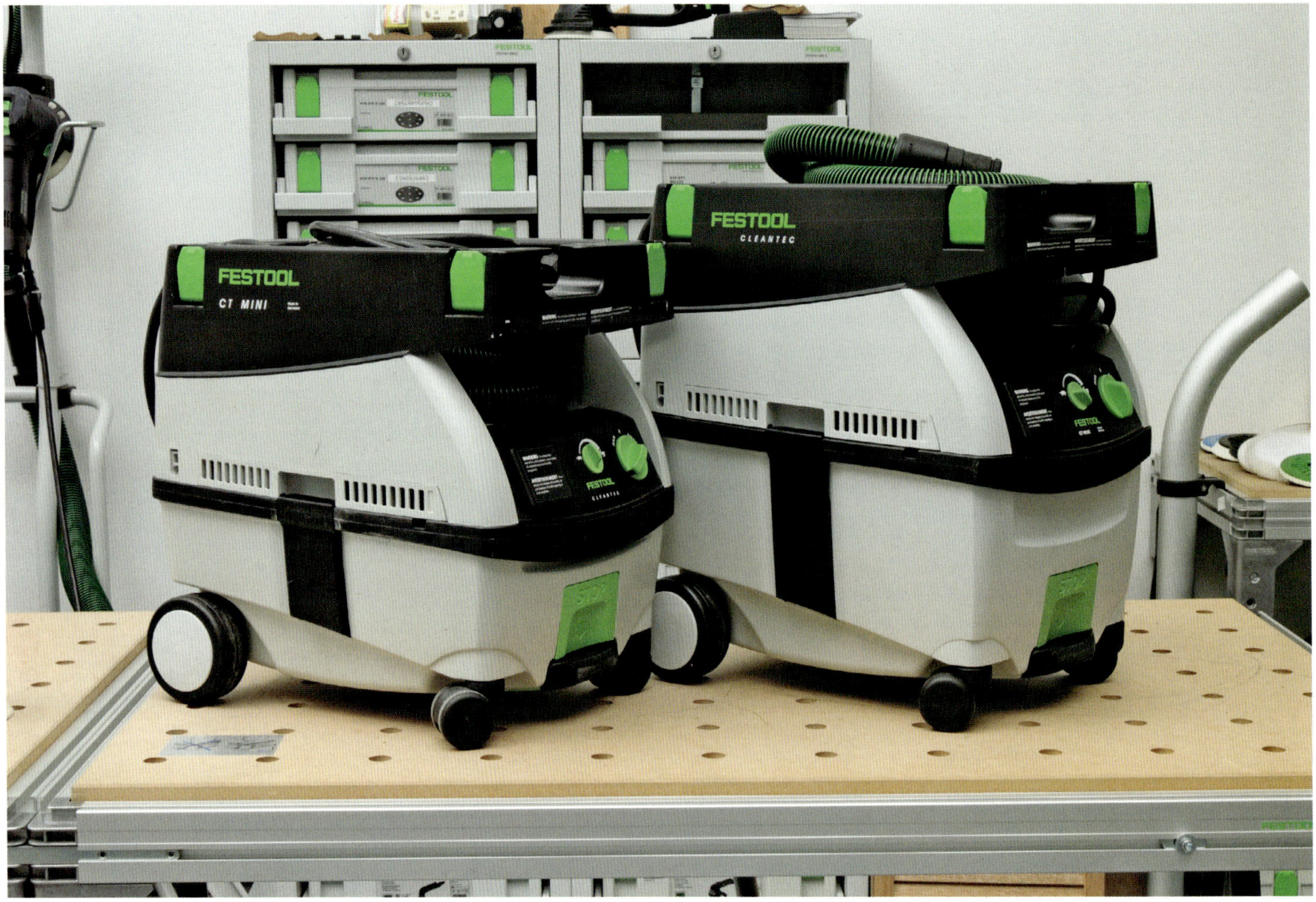

The CT MINI (left) and CT MIDI (right) are nearly identical machines, except for the larger capacity of the MIDI.

Foot brake. When pushed all the way down, it raises the wheels off the ground and keeps the unit in place. This is useful during transport of the unit and on the jobsite.

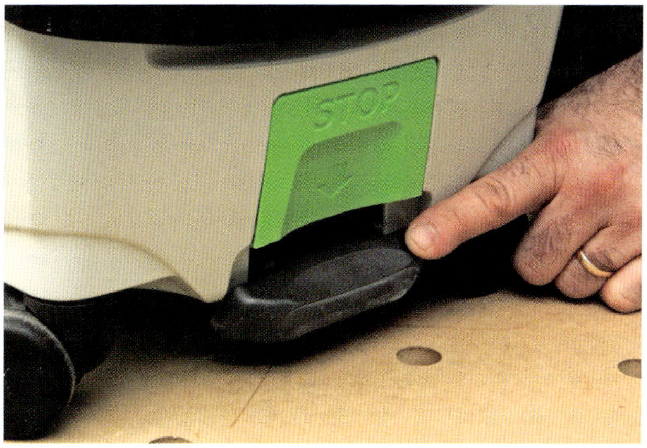

The 27 mm anti-static hose does not allow dust to cling during use. When the vac is turned off, there is no dust in the hose to fall back into the workspace.

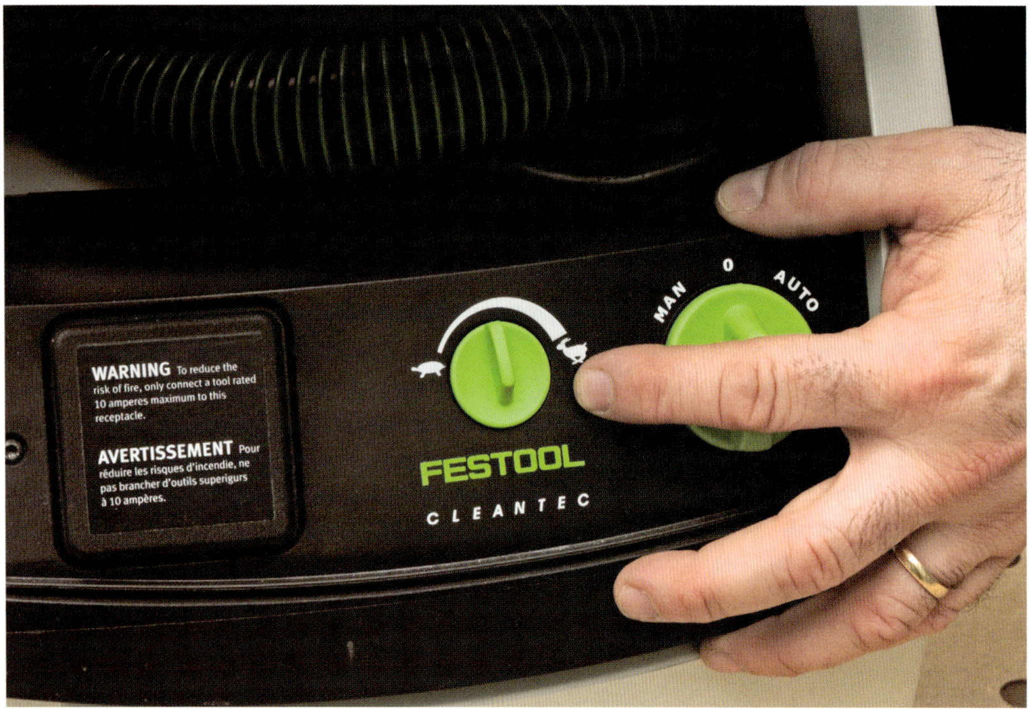

Variable suction control. This is especially important during sanding operations, where a powerful suction could pull the pad too snugly against the work. This would create excessive friction and heat, making swirl marks on the material. By lowering the suction to the correct level, the sanding proceeds much more cleanly.

ON/OFF switch. "0" equals off.

"MAN" is manual, turning the vac on.

"AUTO" is automatic, meaning that the vac is turned on when the tool is on.

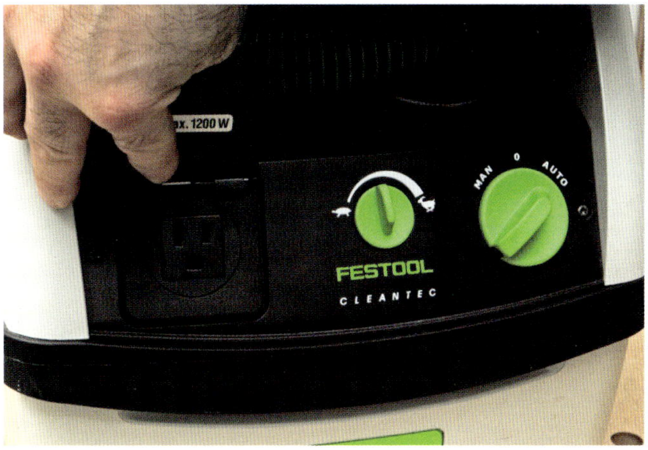

The AUTO function works when the tool is plugged into this outlet on the vac.

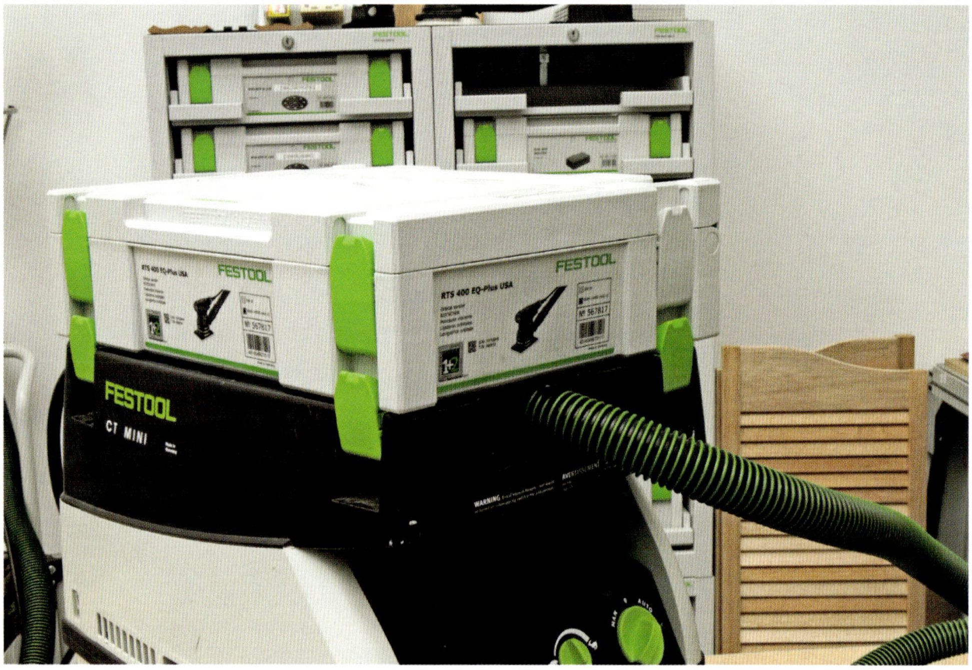

The SYSDOCK allows various SYSTAINERS to be securely attached to the top of the vac for easy transport…the vac actual serves as a dolly.

There are four latches for the SYSDOCK, one at the front and back, and two on one side. With the SYSTAINER in place, open the latch, pull it up and lock it in place.

To gain access to the bag and filters, undo the latches on both sides of the base.

Lift the top by the handle.

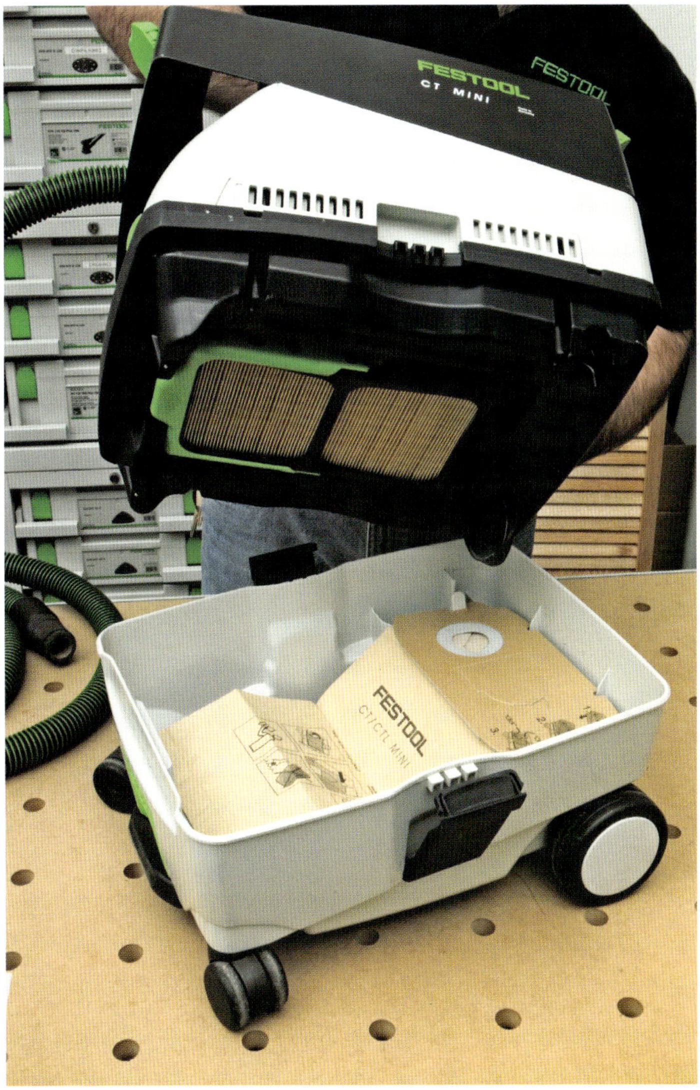

This reveals the filter bag, in the bottom compartment, and the filters under the lid.

The bag in place. The dust comes from the job directly into the bag. Particles 5 microns or greater stay there. The smaller bits move through the bag into the filter, which captures particles 1 micron in size or larger.

Clips on either side hold the bag in place.

When the bag is full, this tab is lifted…

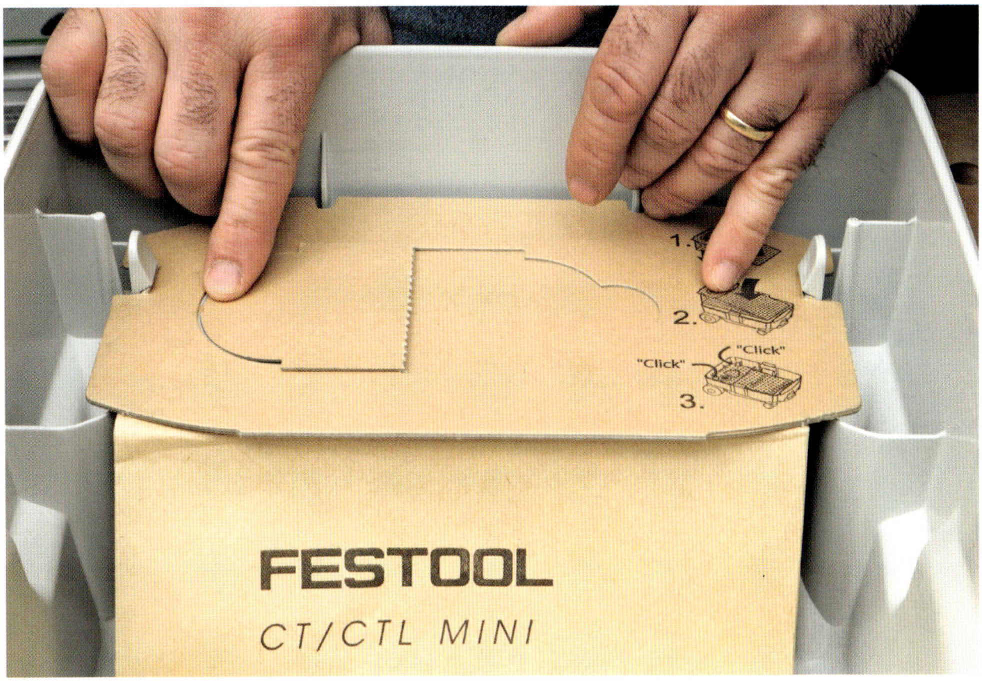

…and pressed in place to seal the bag, keeping the dust inside.

The 1 micron filter.

For cleaning, pull the handle out…

…and remove the filter.

To clean, remove the filter from the frame…

…and tap against the side of a trash can. Used properly the filters will last a very long time. Replace when there is a noticeable loss of air flow, even after replacing the bag and cleaning the air filter.

Extra filter bags can be stored in this space at the back.

The hose and cord are neatly stored in the top.

The vacs come with a 20 amp plug to go into 20 amp outlets, now required in most commercial applications.

The CT MINI & MIDI

Since most residences and older construction do not have these outlets, there is a 15 amp plug adapter available. These are easily lost, so it is best to have a few in your toolbox.

The CT MIDI bag has a 3.3 gallon dry capacity, about 1.3 gallons more than the CT MINI.

For wet vac use, the filter can be replaced with a foam breaking wet filter element.

Remove the bag.

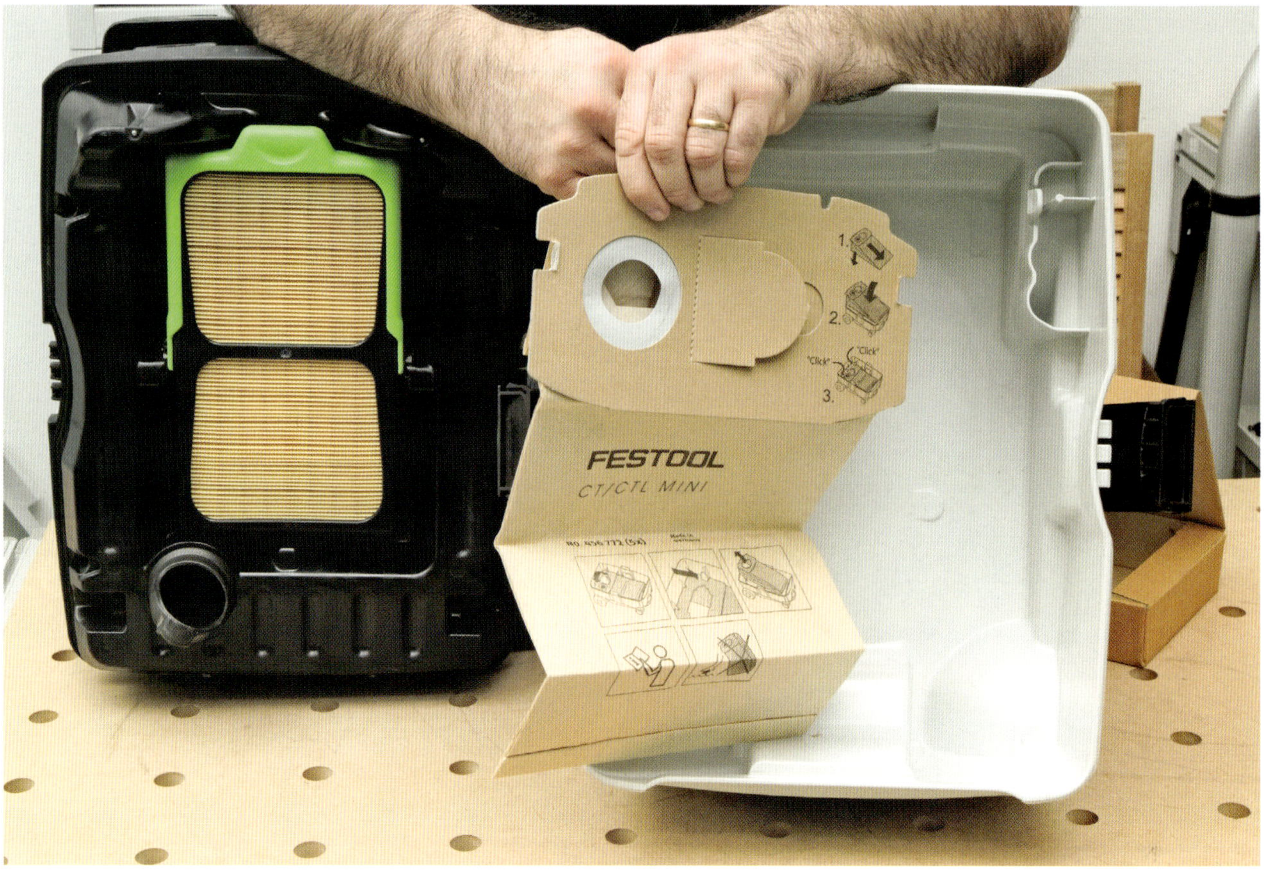

Converting the CT MINI & MIDI for Wet Vacuuming

Remove the 1 micron filter...

...and replace with the foam breaking filter.

There are two water level sensors (level stop sensors) that turn off the motor when the water gets to capacity.

CT 22 E & CT 33 E

The **CT 22 E** and **CT 33 E** are Festool's most powerful mobile dust extractors. With a sturdy chasis, extra large wheels, and double casters, these units can easily negotiate obstacles in the shop or on the worksite. With 134 CFM and only 62 db (on the low setting), these CTs can quickly and quietly remove large amounts of debris, with minimal noise distractions. HEPA filters come standard on the CT 22 E and CT 33 E and offer the ultimate in clean air and filtration. Integrated Sys-Docks or the optional hose garage allow you to attach one or more Systainers for added convenience. An optional boom arm makes a perfect companion in a shop environment and keeps hoses and cords organized and away from work surfaces.

CT 22 E & CT 33 E Features

• **Bypass blower port.** The bypass blower port allows you to evacuate unpleasant odors and fumes from the jobsite, by attaching an additional hose.

• **HEPA filters.** HEPA filters are standard and remove 99.97% of particulate to .3 micron.

• **Automatic tool start and adjustable suction.** Smooth start, infinitely adjustable electronic air flow control and tool-triggered or manual on-off switch ensures convenient and efficient dust control.

• **Convenient carry handle.** Ergonomic handle is positioned 7-9 inches in from the body, which is the most natural position for lifting.

• **Filter scrubber.** The convenient scrubber handle removes the accumulation of dust and debris on the internal filter, without the need to open the machine and improves air circulation across the motor.

• **Sys-Dock and anti-static hose.** The flat top with locking latches allows you to transport tools and accessories securely and conveniently when used with a Festool Systainer. As with all CT dust extractors, the 1-1/16" x 11.5' (27mm x 3.5m) anti-static hose comes standard.

• **Automatic water level sensor and optional dirt trap.** With wet applications, the internal sensors monitor the fluid level and switches the mobile dust extractor off when the container is full (dirt trap available as an accessory).

• **Cord storage.** Everything needs its own storage space. The cord can be wound on the dedicated cord wrap to prevent it from getting in the way during transportation.

• **Tool-less filter changes.** If there is ever a need to change filters, or if you want to switch to a sponge filter for wet applications, changing the filter is very easy. The flat filter can be changed without using tools.

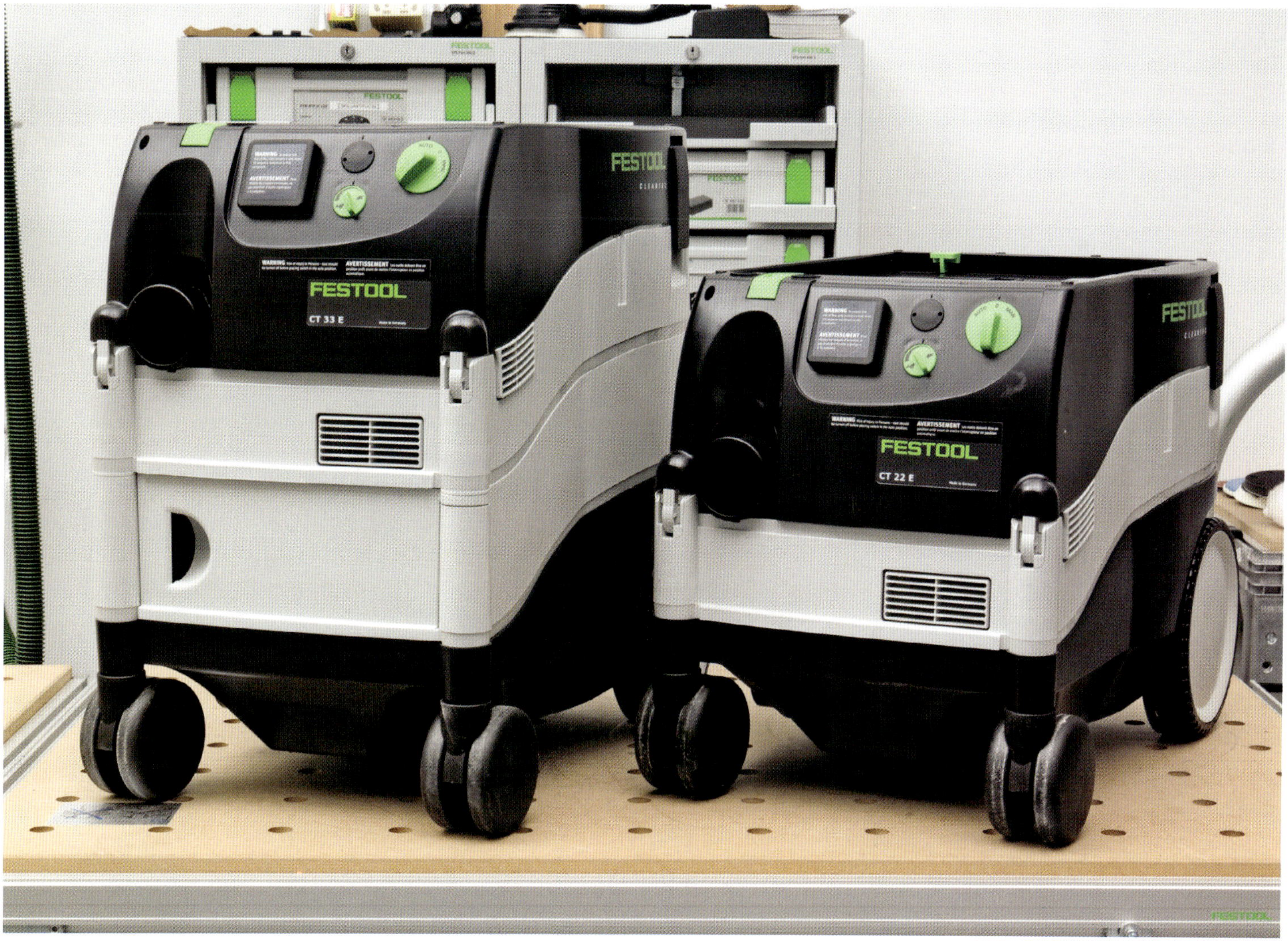

The CT 33 E is nearly identical to the CT 22 E with the exception of capacity. The CT 33 E holds 2.6 gallons more than the CT 22 E.

The CT 33 E has a small storage compartment in front, useful for spare bags.

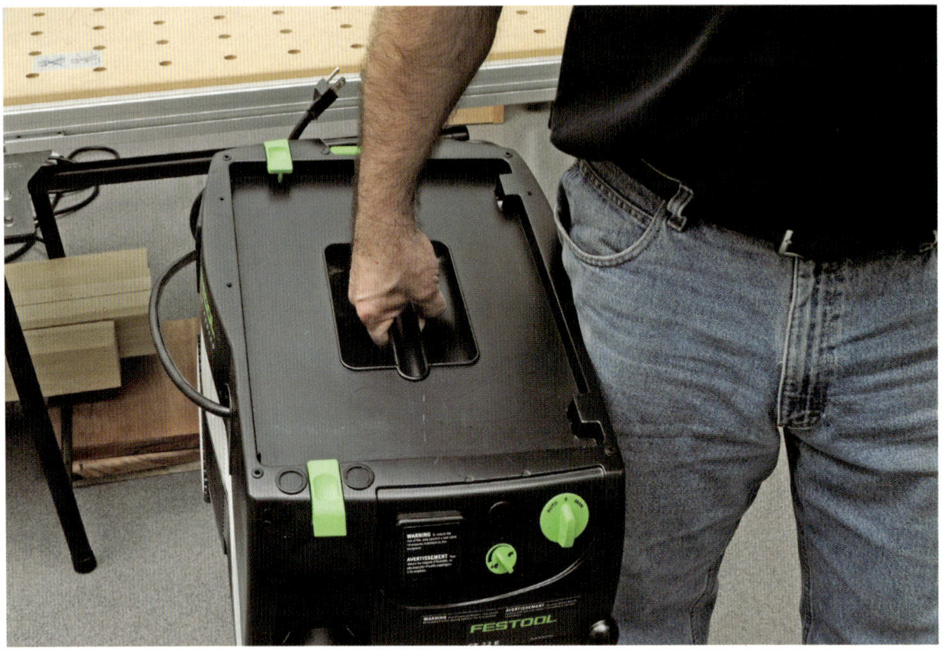

The handle is ergonomically designed to be about 7-9 inches from the body, a comfortable carrying position.

CT 22 E & CT 33 E

The controls are the same as the smaller units…

…with the tool outlet behind the door.

The SYSDOCK system is somewhat different. Tabs at the lower edge of the SYSTAINER…

…fit into sockets on the inner edge of the top of the vac.

Lower the SYSTAINER into place and latch it at the opposite side.

The latch simply slides into place.

Large, non-marring wheels on this unit allow for easy movement.

Wheel locks are located inside the rear wheels.

Simply push them down to engage the lock.

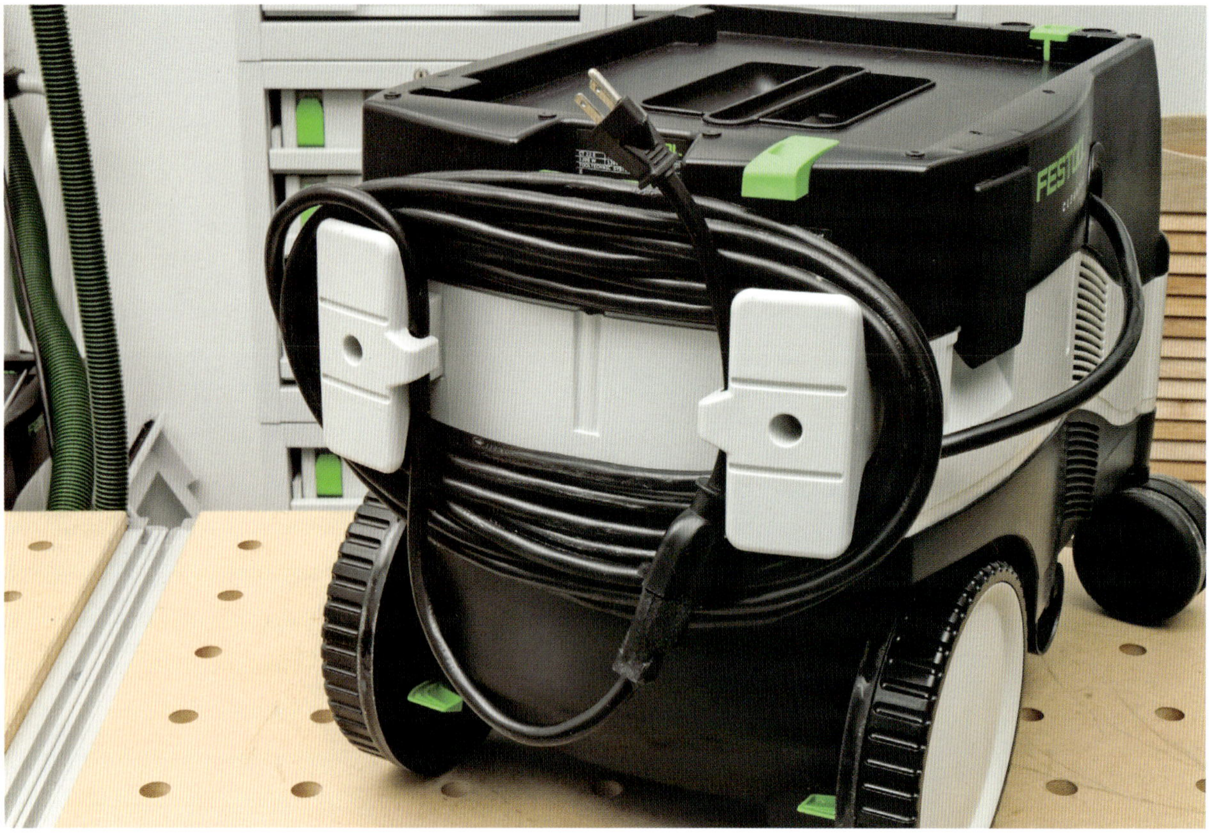

The cord is stored on the rear.

On the side are the exhaust ports. This grille is the normal exhaurst outlet.

Beneath it is the round exhaust port that allows venting to the outside, removing any odors or vapors that are not captured by the HEPA filter.

There is a flap inside that raises when the hose is inserted, blocking the normal venting system.

The exhaust hose can be run to a place outside the work area.

The green handle at the back of the unit operates the quick filter scrubber.

Pull it in and out a couple of times while the unit is off and it knocks the dust into the reservoir. This will prevent the filters from caking. A thorough cleaning is done by removing the filters and tapping them.

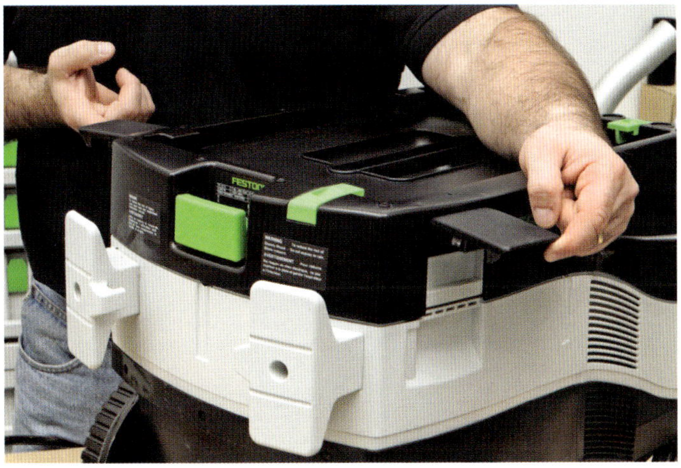

To access the bag and filters, lift the release latches on both sides.

The lids lifts from the back.

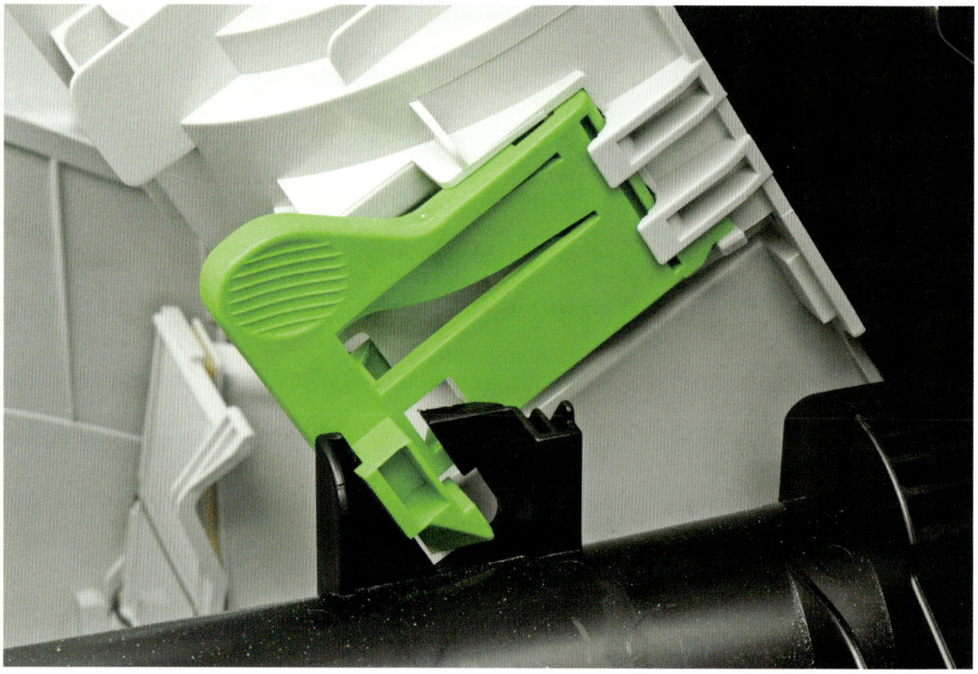

A kickstand automatically engages to keep the lid up.

To release the kickstand, simple press in.

The 5 micron bag fits in the well.

To remove it, pull back on the tab at the opening.

Insert the cap to seal the bag.

Insert a new bag, using your hand to keep it straight.

When it is snuggly in place, bend the tab down. This promotes uniform filling and prevents tearing when removing the bag.

This is the grounding contact for the hose. The grounding strap allows the unit to have continuity for the anti-static hose design and ground the intake port to the electrical ground.

The latch for the HEPA filter.

Depress the latch and pull down the filter frame.

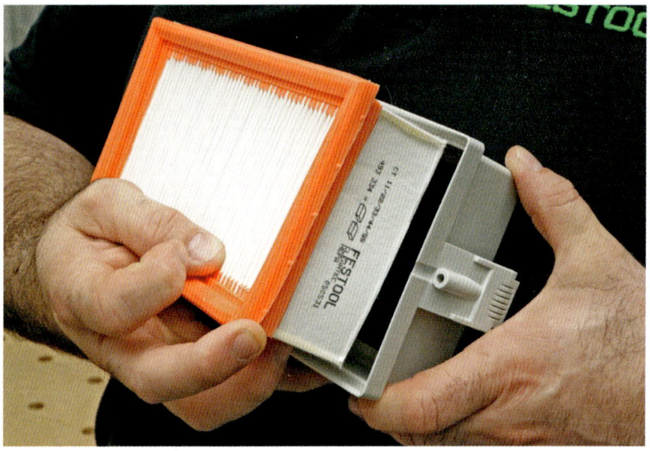

When the frame is free from the unit, the filter simply pops out.

Tabs at the front of the filter frame fit into slots in the unit.

Insert them and pivot the frame into position. The black apparatus above the filter tray is the quick filter scrubber.

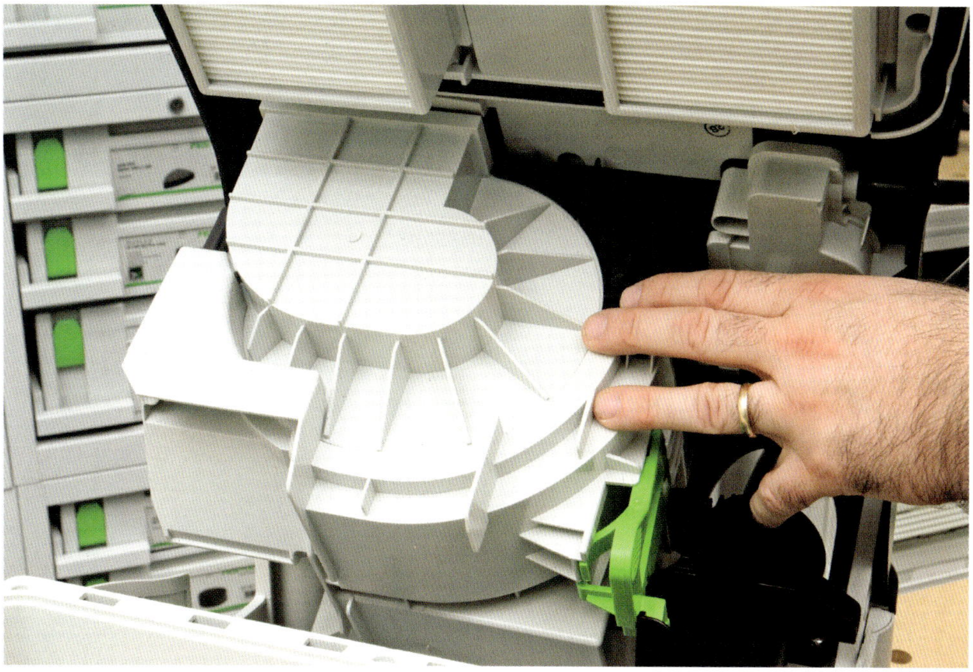

The motor is completely encased, keeping it free of dust and insuring a long life.

In wet vac mode, the level stop sensors turn the motor off when the reservoir is filled.

Wet Vacuuming with the CT 22 E & CT 33 E

To convert to wet vac mode, remove the bag.

Remove the filter…

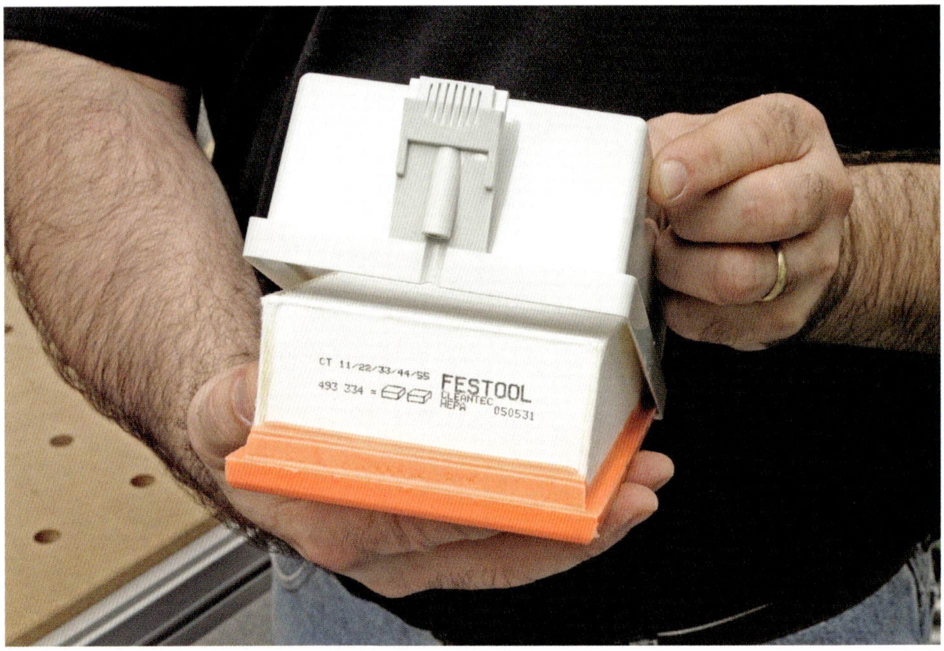

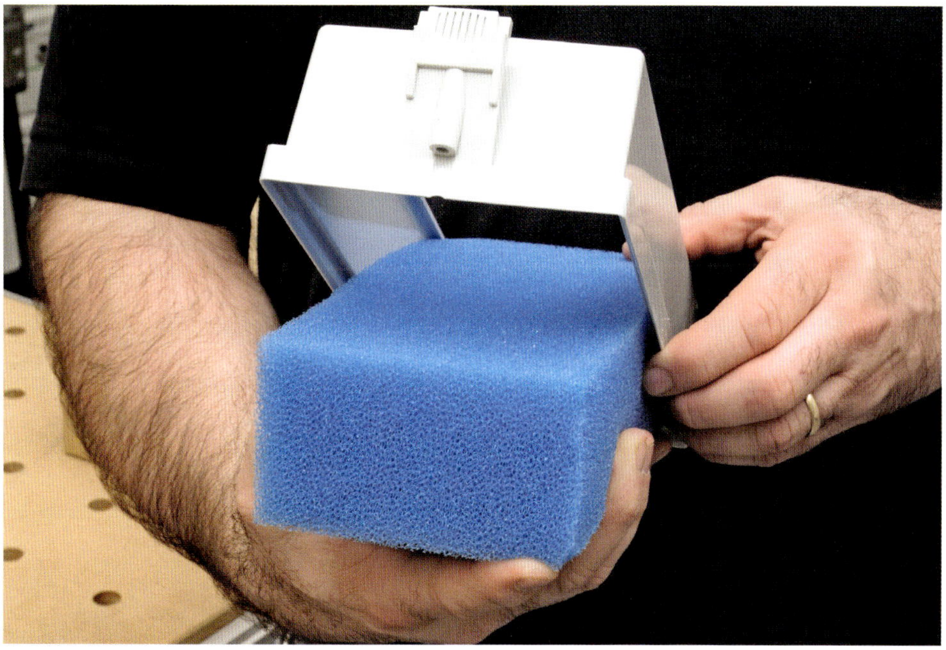

...and replace with foam breaking wet filter element.

Reinsert the filter frames.

Wet Vacuuming with the CT 22 E & CT 33 E

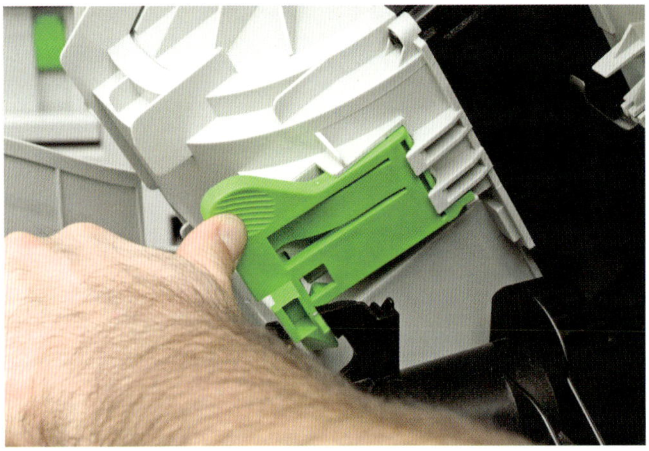

When the reservoir is full you can remove the top to make emptying it easier. Release the green kickstand.

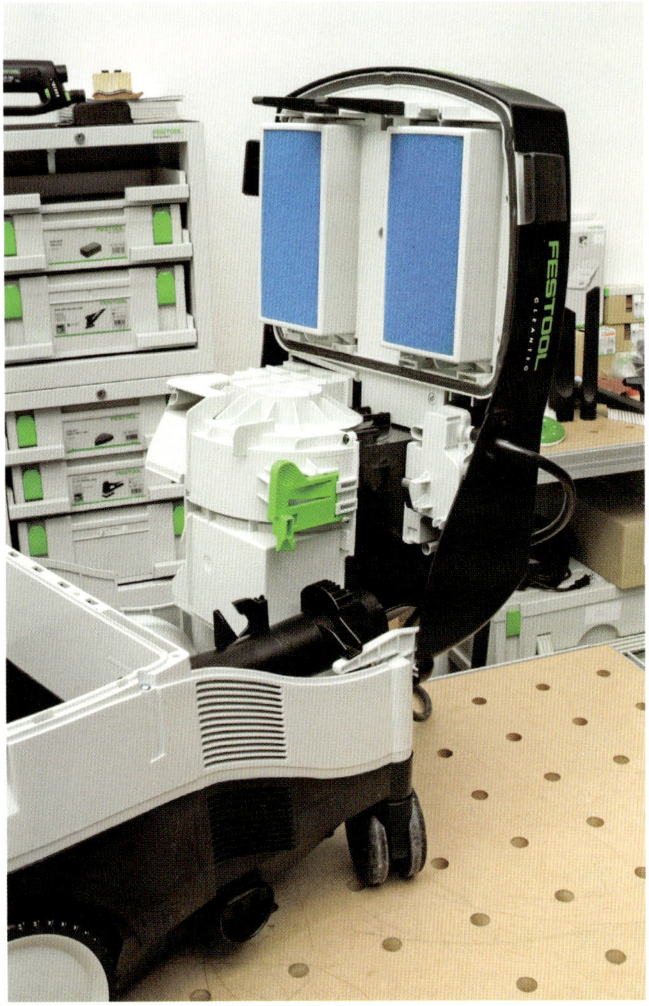

Pivot the top to the vertical position.

Pull the top straight out.

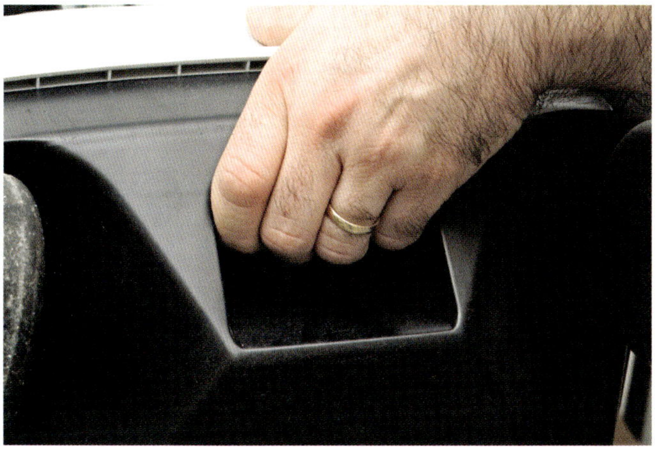

There is a handle at the front end…

…which makes emptying quite easy.

Wet Vacuuming with the CT 22 E & CT 33 E

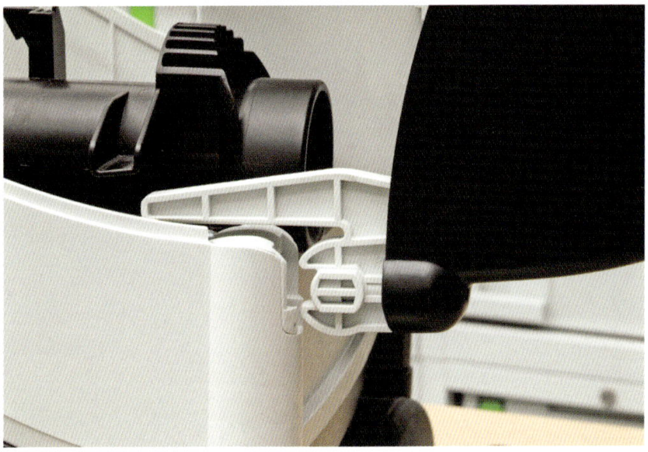

Reinsert the top. The long piece on the left goes above and the smaller rounder tab gets inserted into the slot.

Progress.

In place.

Wet Vacuuming with the CT 22 E & CT 33 E

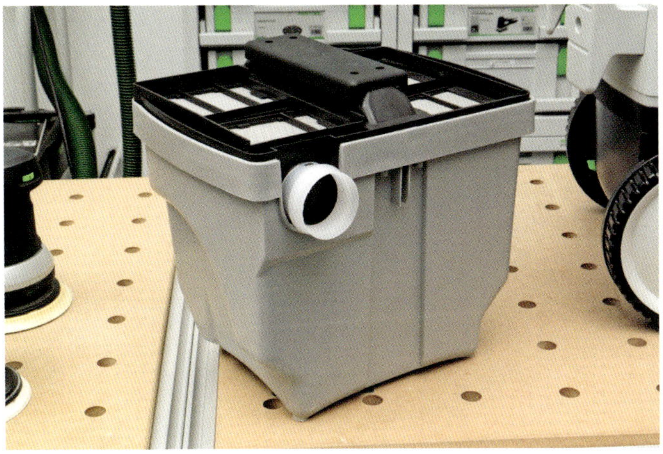

The Dirt Trap accessory provides an alternative for wet vac applications.

Insert the pail into the well.

The dirt screen then fits on top. Align the input in the hole...

...and lower into place. The pail can simply be removed for emptying.

Hoses

The standard hose for all dust extractors is antistatic, 27mm (1-1/16 inch) diameter, 3.5 meter (11.5 feet) long. This coordinates with the standard tool power cord, which is 13 feet long. It is also available in lengths up to 16.5 feet, which would require an extension cord.

A 36mm (1-7/16 inch) antistatic hose is available for the CT 22 E and CT 33 E. This will prevent clogging when using tools that produce larger chips and sawdust quickly, like a router or planer. Available in lengths of 11.5 feet, 16.5, and 23 feet.

The dust shrouds of the tools are designed to fit with both the 27mm and the 36mm hoses. On the 27mm, the hose fits in the shroud.

On the 36mm, the shroud fits in the hose.

Hoses

The largest antistatic hose is 50mm (2-inches). It is offered in 8.25 and 13 feet. Its main purposes are to connect the dust extractor to shop tools and for shop clean-up.

All the hoses just discussed are available as non-antistatic, "gray" hose. This is good for the exhaust feature on the CT 22 E and CT 33 E, and for connection to a dust extractor by other manufacturers that do not offer the antistatic feature.

Connectors for the CT 22 E and CT 33 E

The Y-piece, allows the dust extractor to be connected to two devices. It plugs into the extractor, and has a blanking plug to close one of the branches when not in use.

Sometimes it is necessary to extend the hose. The combination depends on the application. In this case a 36mm hose from the extractor, on the right, is connecting to a 27mm hose going to the tool, on the left.

Connectors for the CT 22 E and CT 33 E

First, pull off the collar.

Then, unscrew the 36mm rotating connector collar. It follows the ribs in the hose, so it unscrews in a clockwise direction.

The rotating connector comes in antistatic and non-antistatic varieties. It has three parts as shown here. Left to right: rotating hose connector, threaded hose sleeve, and hose connector clip.

Connectors for the CT 22 E and CT 33 E

Place the clip around the hose.

Screw the hose sleeve onto the hose, turning it counterclockwise.

Clip the rotating connector in place.

Connectors for the CT 22 E and CT 33 E

Complete.

The two hoses can now be joined with the connector sleeve or to a hose reducer coupling sleeve as needed.

Connectors, left to right: Antistatic connector sleeve; antistatic rotating connector sleeve for 27mm suction hose; antistatic rotating connector sleeve for 36mm suction hose; antistatic rotating connector sleeve for 50mm suction hose.

A blanking plug can be used to seal the intake hole in the dust extractor. The pointed tail…

…fits into a slot beneath the hole.

This is a convenient accessory for keeping the dust from escaping during transport.

The spark trap can be added to the system and is vital when working with metal. The baffled chamber cools down the metal particulate before it hits the bag. It helps to reduce the chance of fire from stray sparks while maintaining air flow.

The hose holder is inserted into the accessory socket at the upper front corner

Other Accessories

The hose holder is inserted into the accessory socket at the upper front corner

The hose holder gets the hose up and level with the work area.

Example.

The hose hook fits on the front on the CT 22 E or CT 33 E, with the rod inserted in the socket at the front corner, and a screw holding it in place.

It provides convenient hose storage.

Other Accessories

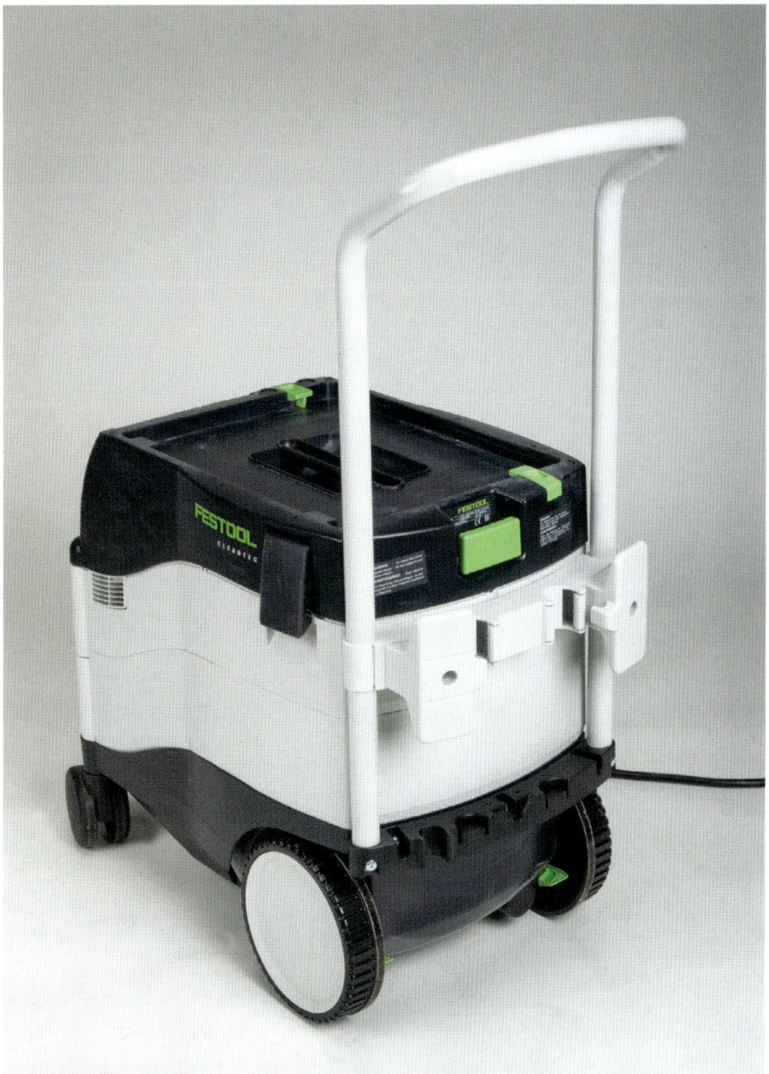

The optional handle assembly makes moving the CT 22 E or CT 33 E even easier. With it the unit can be pushed while standing as opposed to leaning over or carrying. When transporting several Systainers on the SysDock feature, the handle converts the CT into a useful "dolly."

The installation of the optional handle begins with the removal of the cord holders from the back of the machine.

Install the bottom handle assembly…

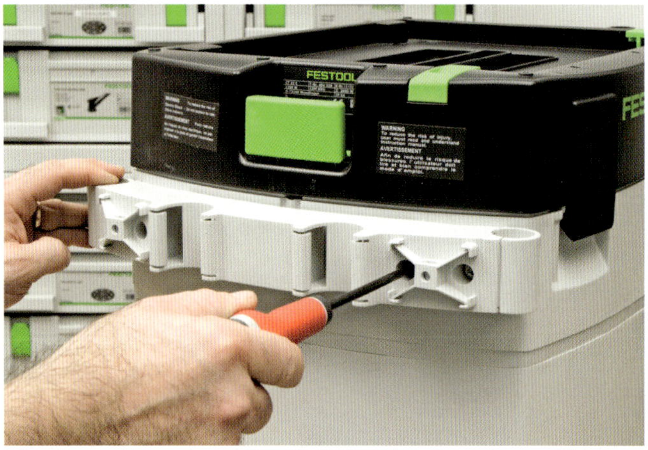

...and the upper handle assembly.

Insert a nut into the slot.

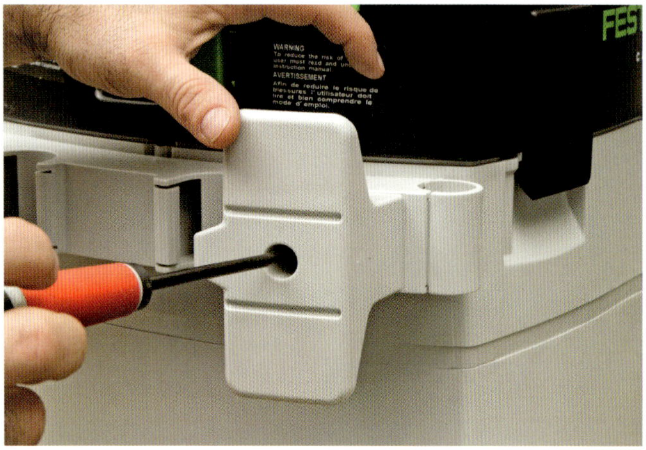

Reinstall the cord holders.

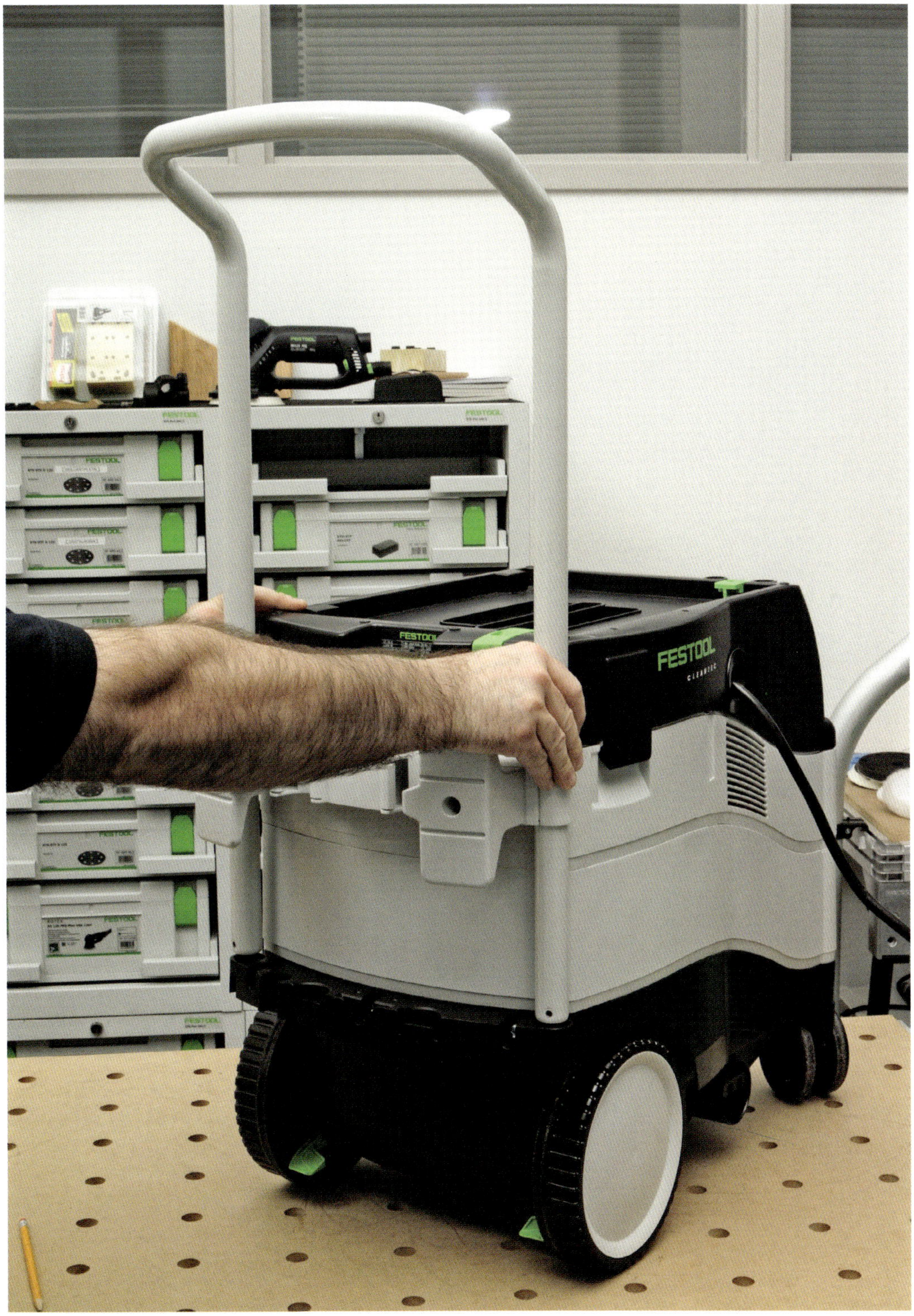

Insert the handle…

...and secure.

The CT hose garage fits on the top and provides convenient storage space for the hose and cord. Set it in place, with tabs in the slots along the left side…

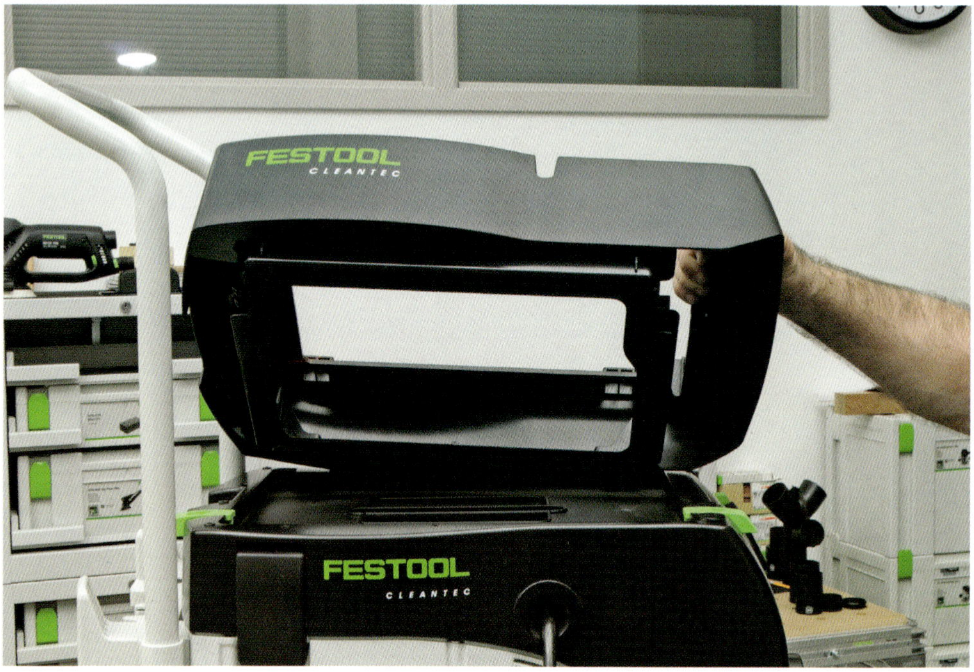

...and lock it.

The Longlife filter bag provides a reusable alternative to paper.

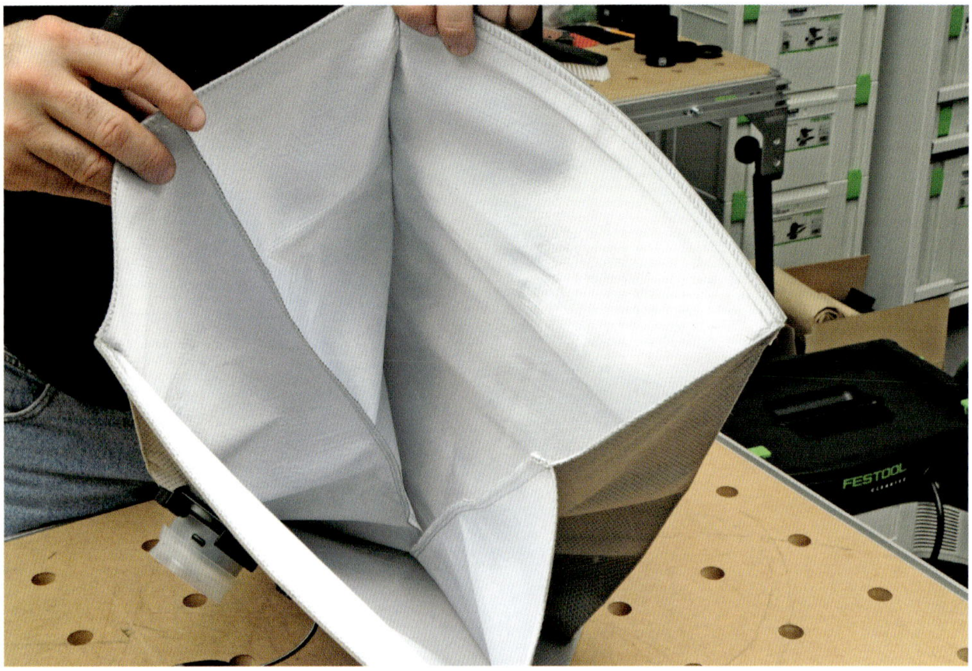

It is lined with fabric for fine filtration.

It closes with this slider.

Place the bag in the well and connect to the dust extraction tube.

The Boom Arm for the CT 22 E and CT 33 E

The Boom Arm accessory keeps the dust extractor hose accessible to the work area, while lifting it up and out of the way of the work at hand. The pivoting arm moves with the tool so it doesn't hinder the work process. It works with the CT 22 E and the CT 33 E and requires the handle assembly. The boom arm comes with a 50mm antistatic extension hose and 3-wire 12 gauge extension power cord. It will accommodate a 27mm or 36mm tool hose, as well as the Festool pneumatic 3-in-1 hose for air tools.

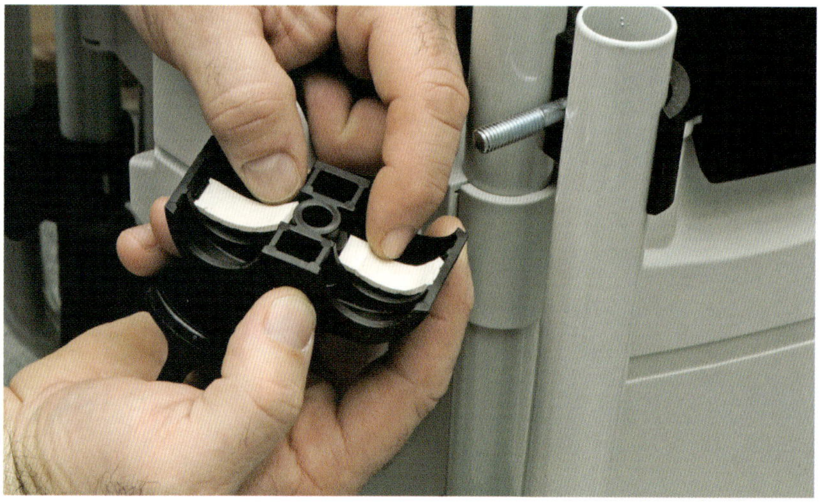

The boom arm requires the handle assembly to be in place. Attaching the boom begins with the outriggers, two added feet that give the unit stability. They are attached to the handles with two tool clamps on each side. The white foam rubber protects the handle assembly from damage.

The Boom Arm for the CT 22 E and CT 33 E

With the bolt side of the clamp in position, place the outrigger tube and the other side of the clamp over the bolt.

Secure with the hand nut.

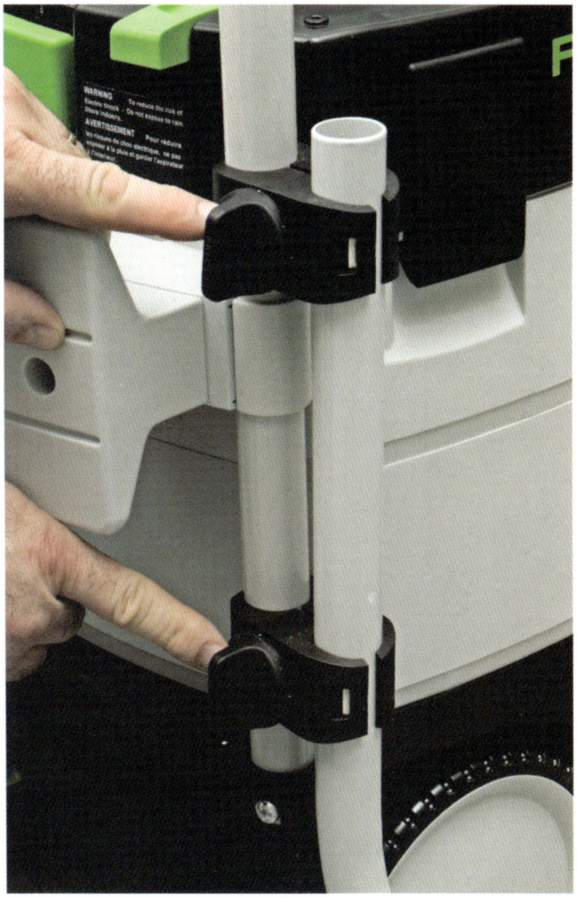

Each side has two tool clamps.

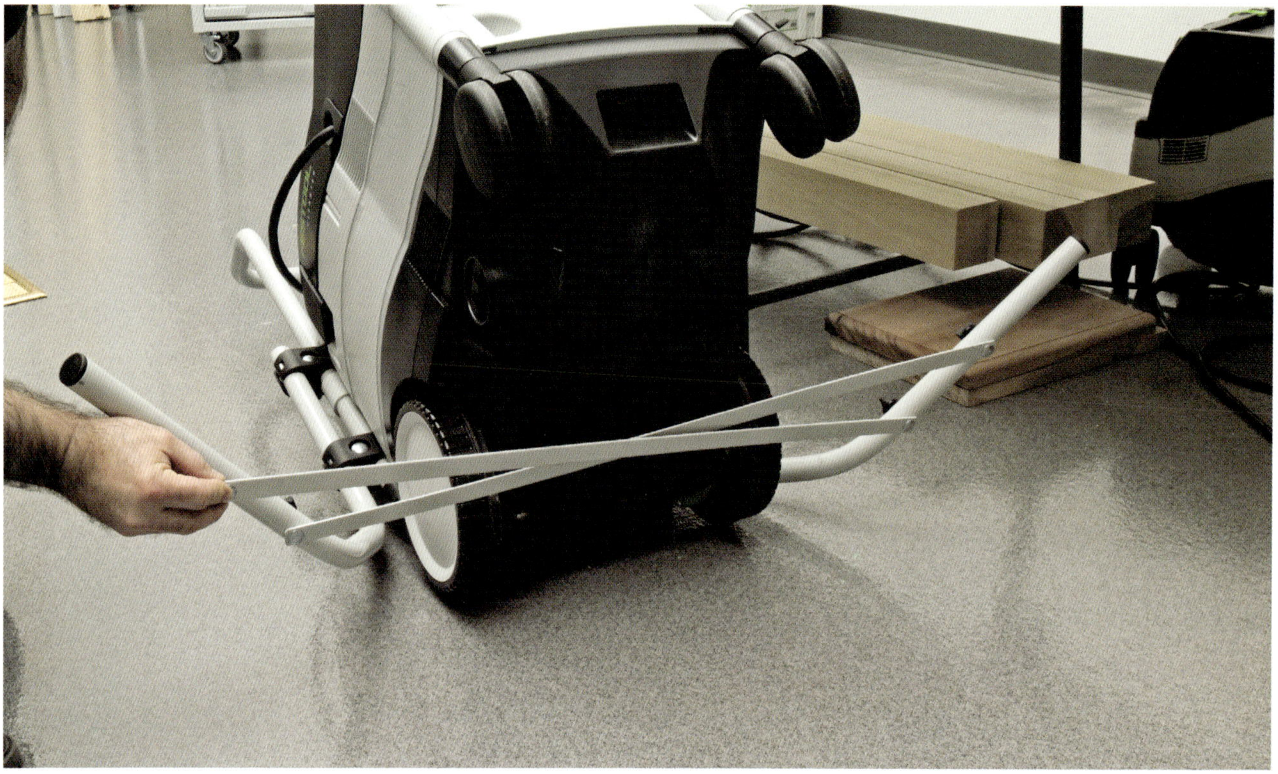

The outriggers are held in position by cross braces that run under the extractor.

Bolt in place, tightening from above. The braces are designed so they do not interfere with the wheels.

The Boom Arm for the CT 22 E and CT 33 E

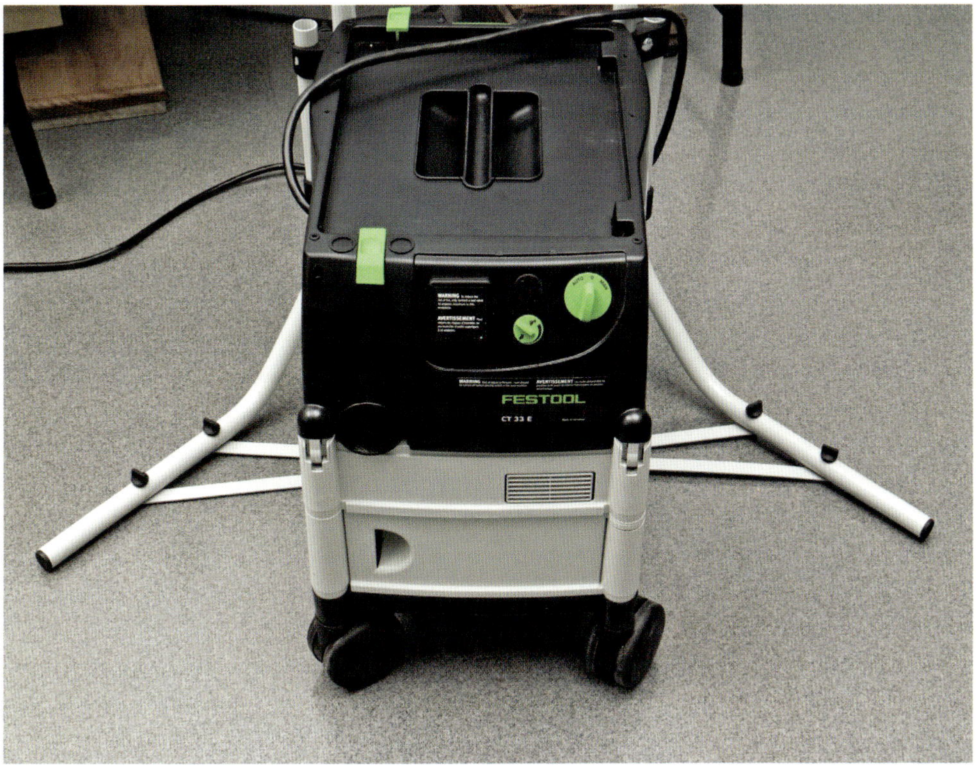

The outriggers in place.

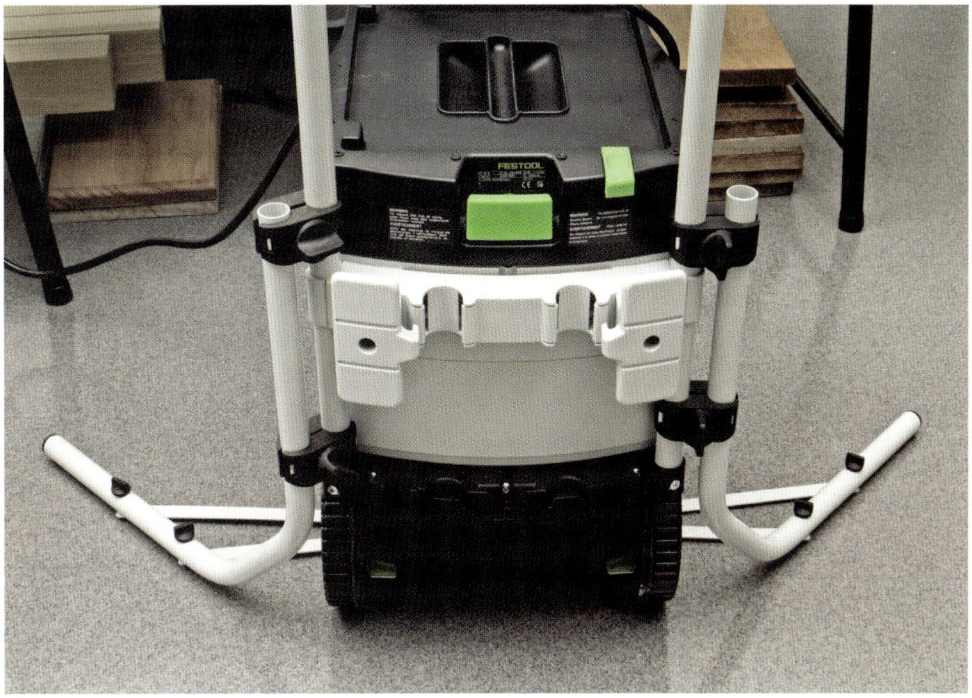

The bottom of the vertical support has a tab. The goes outside of the lower handle bracket to prevent rotation of the tube.

Slide a screw clamp over the vertical support.

The Boom Arm for the CT 22 E and CT 33 E

Place the support in its slot on the lower handle bracket, tab side out. The screw clamp should be below the upper handle bracket.

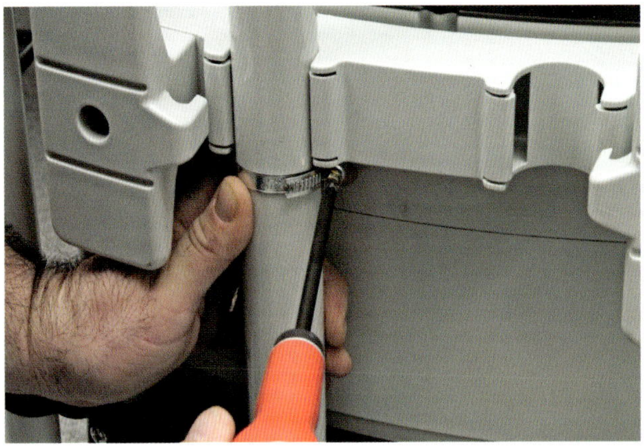

Tighten the screw clamp so that it is snug against the underside of the upper handle bracket. This will act as a stop to help prevent the vertical support from pulling out.

The vertical support in place.

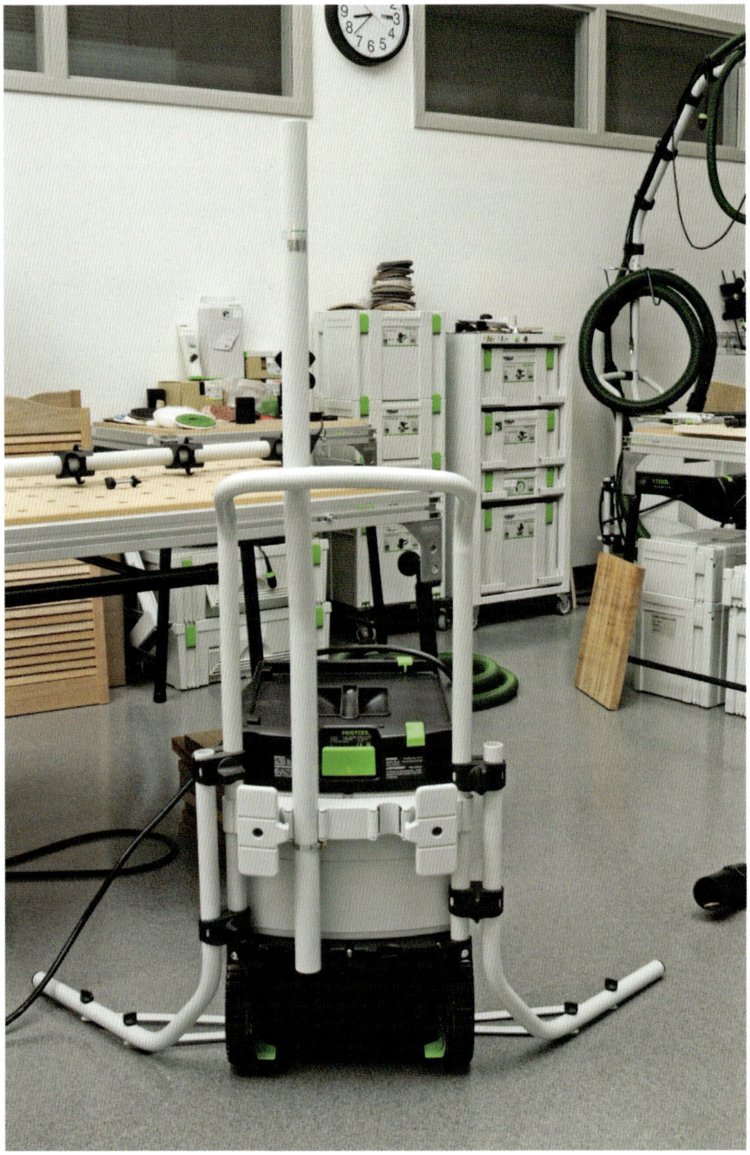

The Boom Arm for the CT 22 E and CT 33 E

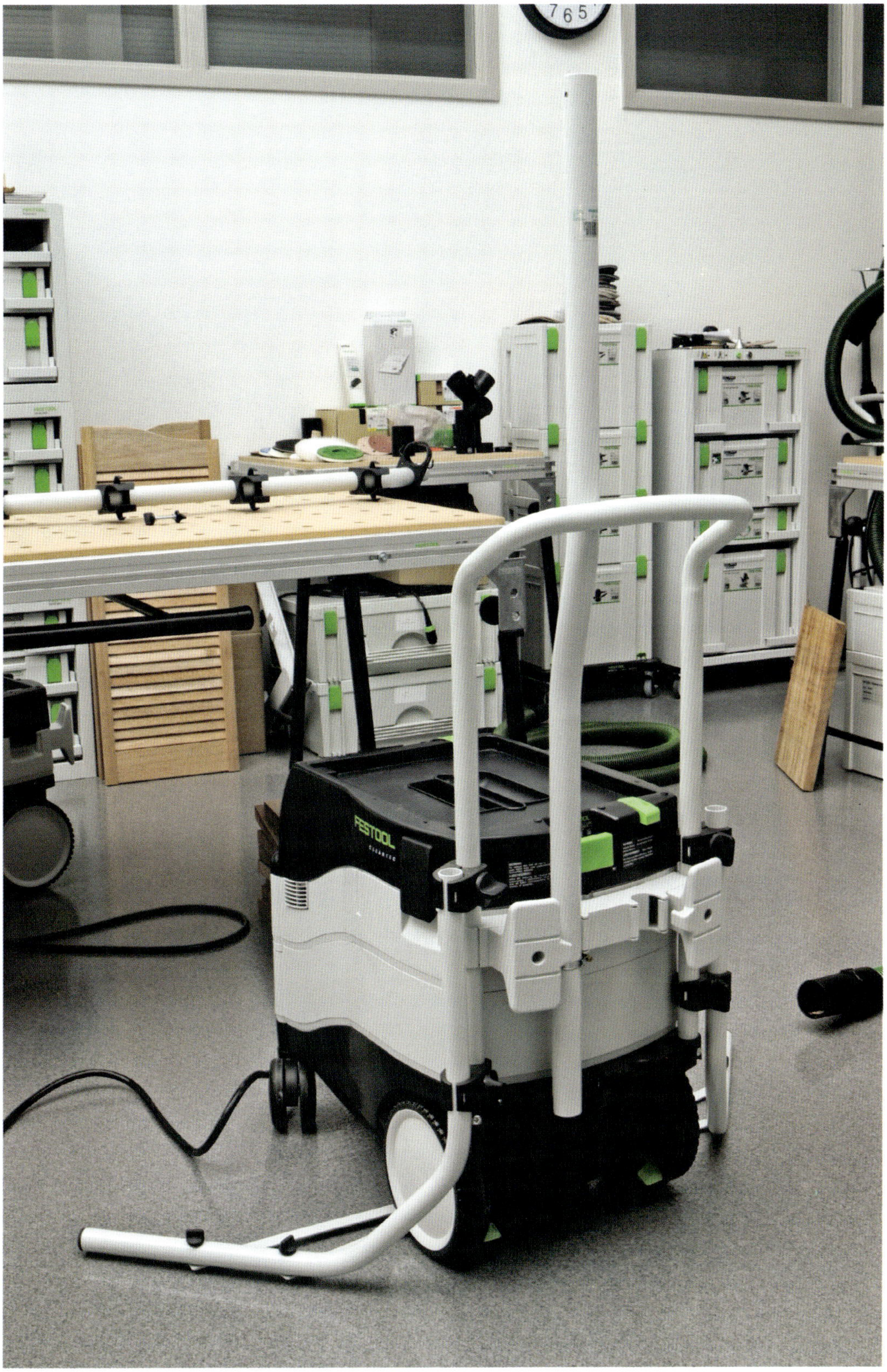

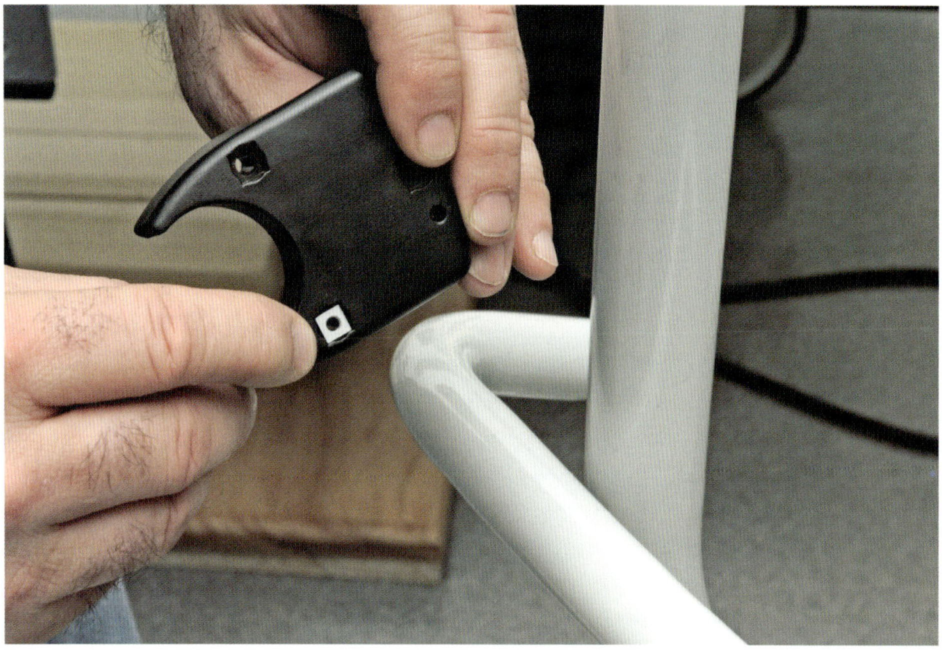

The vertical support is held to the handle with a clamp. The underside holds three recessed nuts.

Place the bottom piece around the vertical support and under the handle.

The Boom Arm for the CT 22 E and CT 33 E

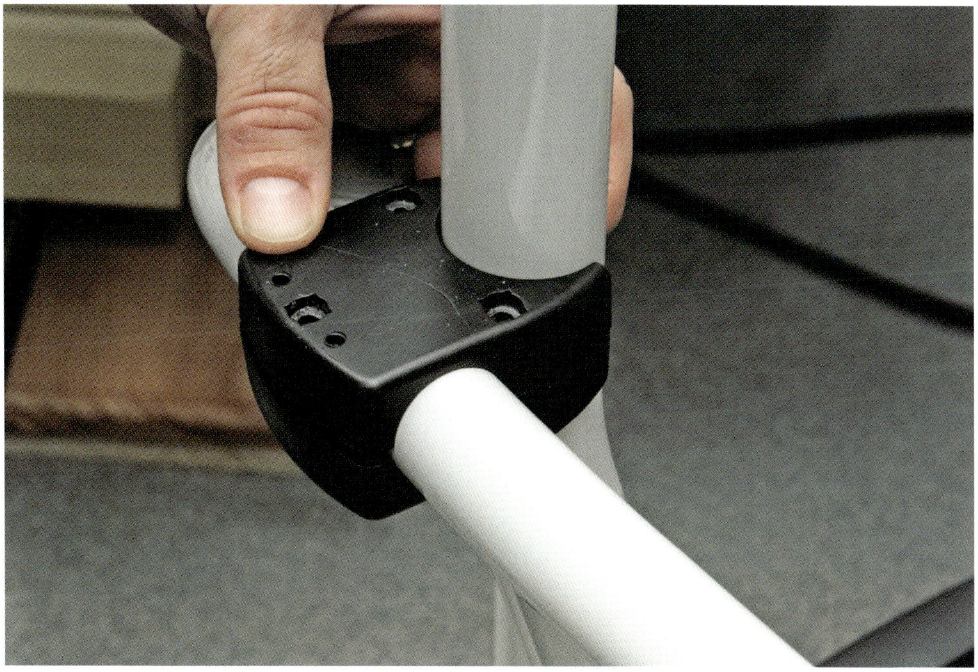

Place the top piece in position.

Hold it in place with three bolts.

The Boom Arm for the CT 22 E and CT 33 E

The vertical support is secured to the bracket with a wire clamp, which is inserted into two holes at the back of the bracket.

In place.

The Boom Arm for the CT 22 E and CT 33 E

The arm arrives in two segments.

Snap them together.

The Boom Arm for the CT 22 E and CT 33 E

Insert the arm into the vertical support.

Secure with the bolt provided.

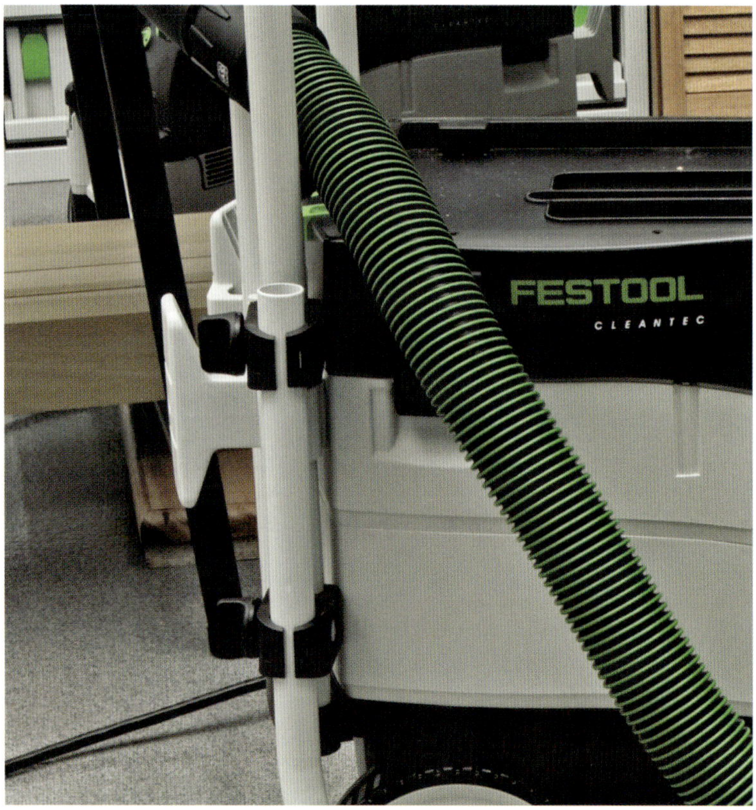

Ready for installation of the hose and power cord.

The Boom Arm for the CT 22 E and CT 33 E

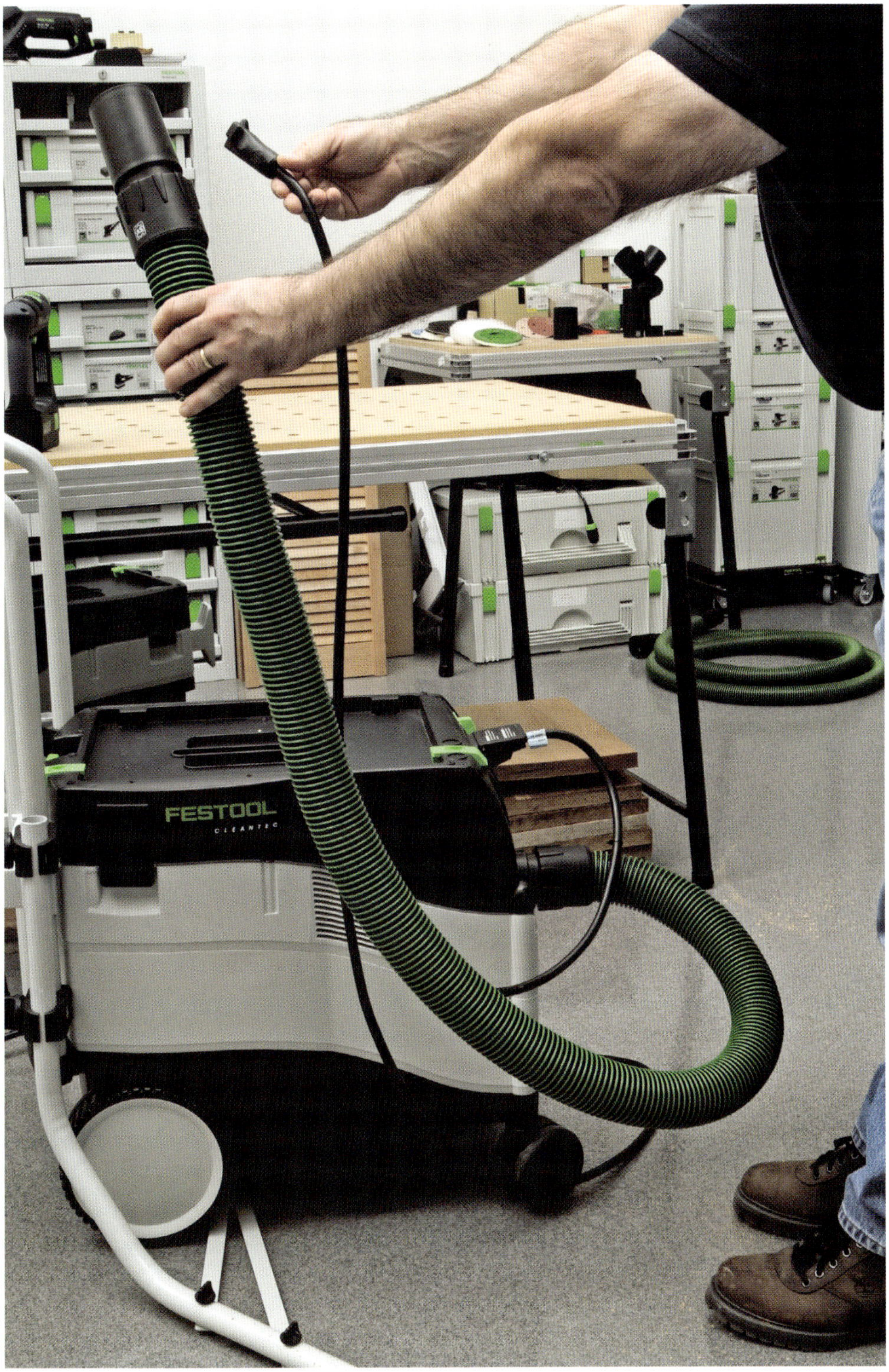

The arm comes with a 50mm antistatic extension hose and 3-wire 12 gauge extension power cord, that run from the extractor to the boom arm.

The arm will accommodate a 27mm or 36mm tool hose, as well as the Festool pneumatic 3-in-1 hose for air tools. Insert it in the hose brackets, starting where the arm joins the hose coming off the dust extractor.

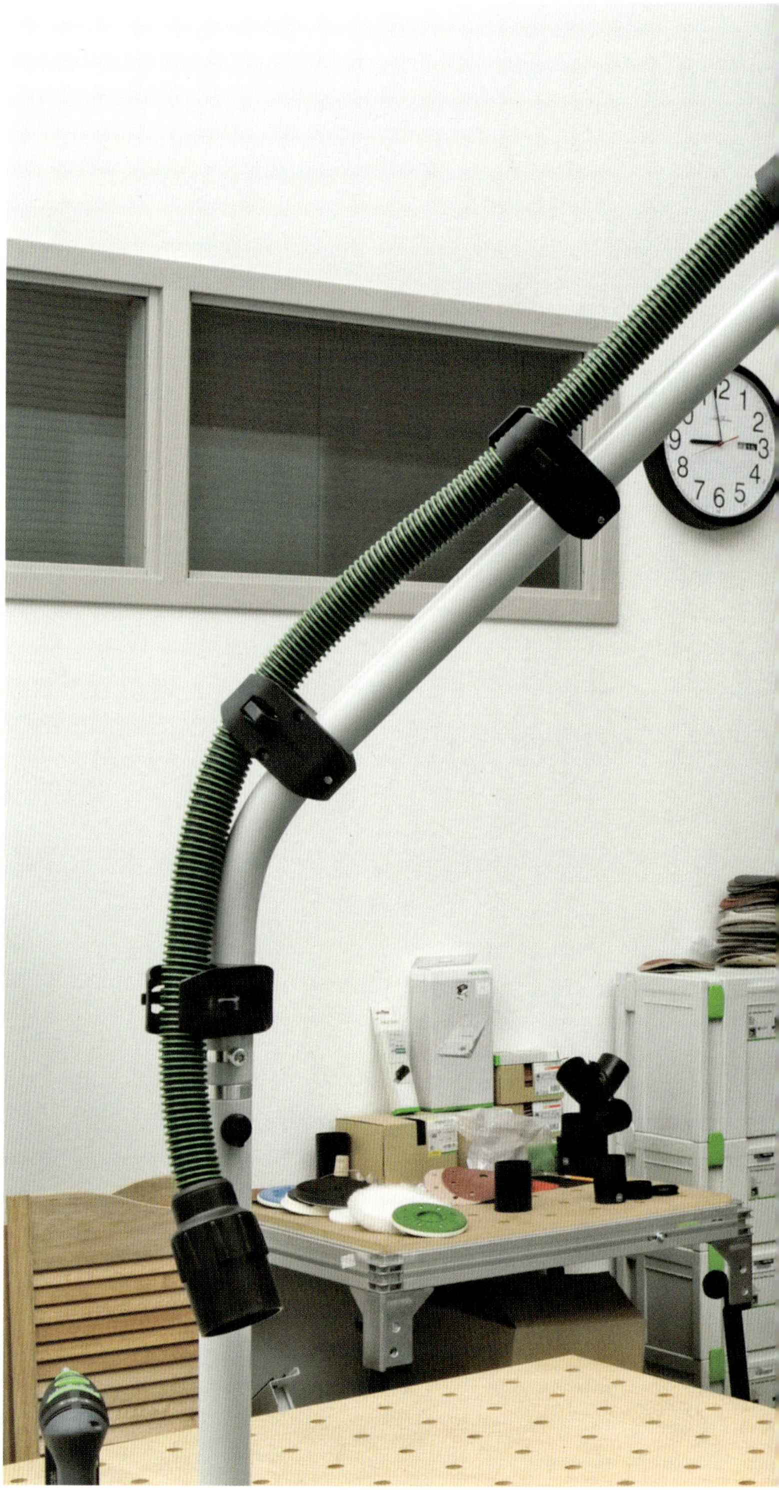

Continue to the end of the arm

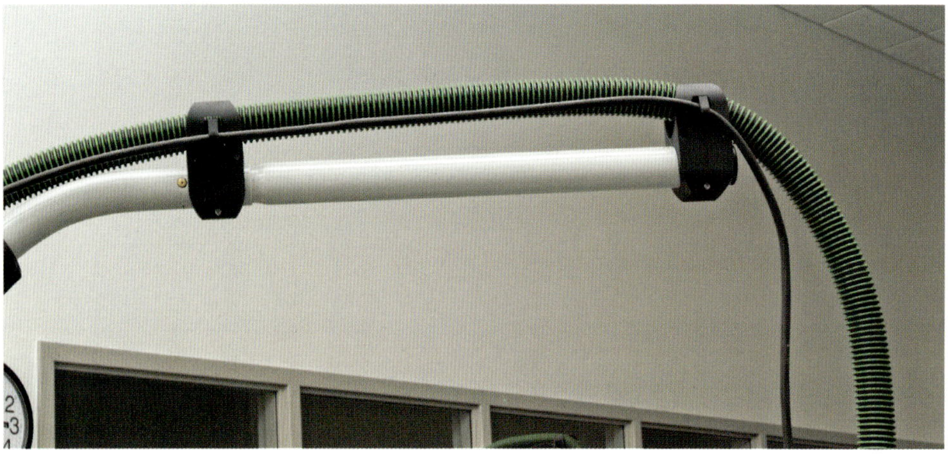

The tool's power cord is clamped to the outside of the hose brackets.

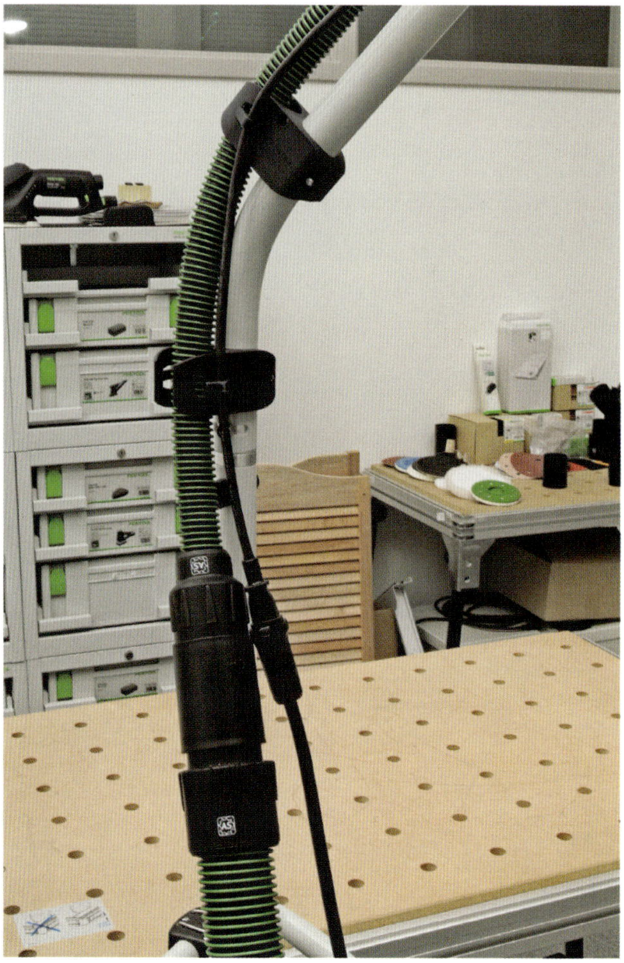

Connect the power cords and hoses.

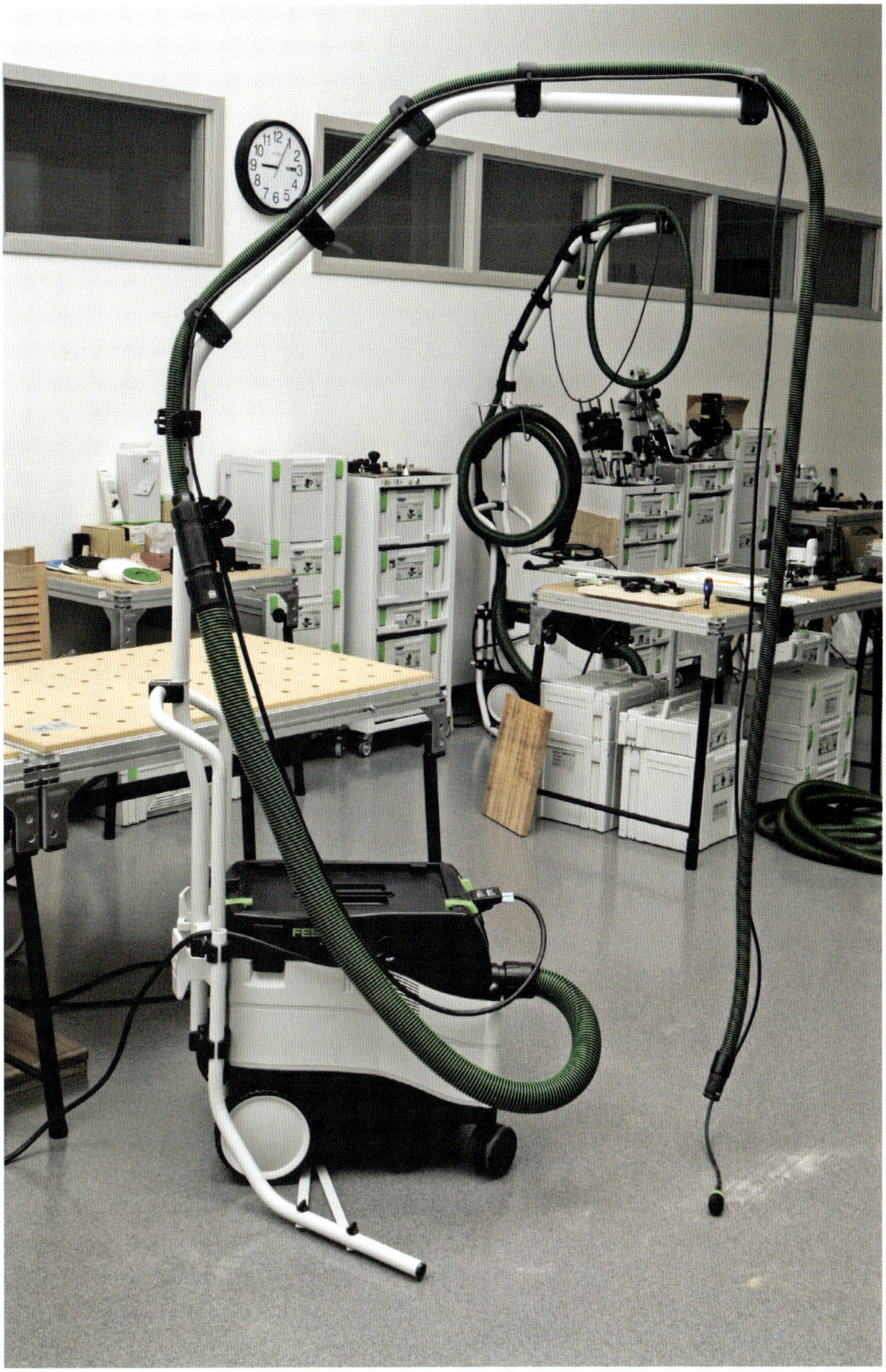

Complete

Tubes, Nozzles, & Brushes

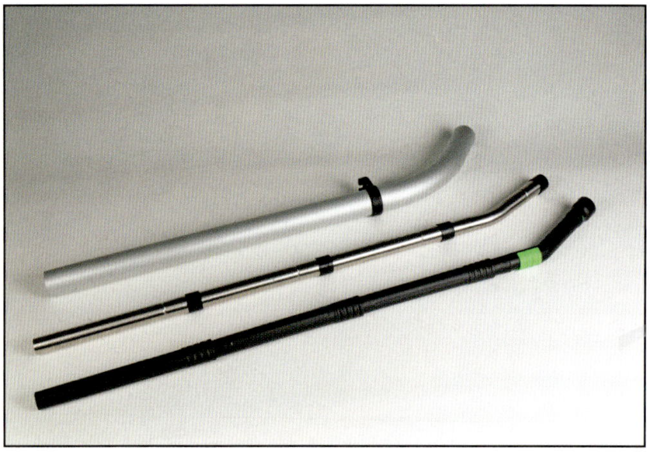

Top to bottom: one-piece anodized aluminum tub; stainless steel curved tube and extensions; curved polyproplyene (PP) curved tube with regulating slide to adjust suction power.

Nozzles & brushes (left to right): Large crevice nozzle; crevice nozzle; universal brush nozzle; upholstery brush; bevel ended nozzle; suction brush.

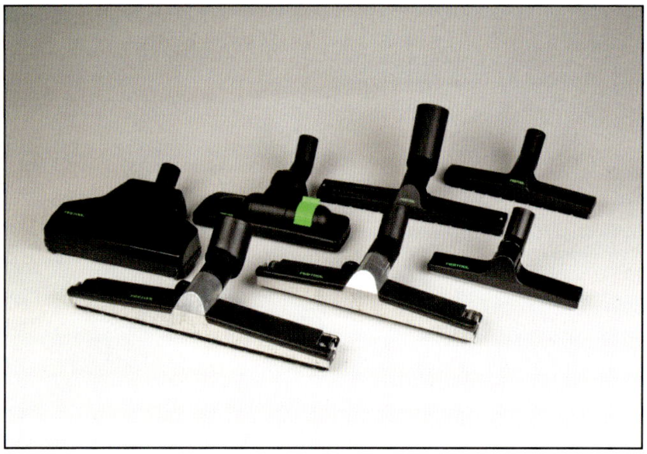

Floor nozzles (front row, left to right): Large industrial floor nozzle; industrial floor nozzle; interchangeable floor nozzle; (back row, left to right): turbo suction brush; multi-purpose floor nozzle; workshop floor nozzle; standard floor nozzle.